Caregivers of Adults with Intellectual Disabilities:

The Relationship of Compound Caregiving and Reciprocity to Quality of Life

by

Elizabeth A. Perkins

A dissertation submitted in partial fulfillment
of the requirements for the degree of
Doctor of Philosophy
School of Aging Studies
College of Behavioral and Community Sciences
University of South Florida

Major Professor: William E. Haley, Ph.D
Matthew P. Janicki, Ph.D
Cathy L. McEvoy, Ph.D
Sandra L. Reynolds, Ph.D
Brent J. Small, Ph.D

Date of Approval:
November 4, 2009

Keywords: developmental disabilities, aging, caregiving, parents, family

UMI Number: 3420612

ProQuest LLC
789 East Eisenhower Parkway
P.O. Box 1346
Ann Arbor, MI 48106-1346

Dedication

I dedicate this dissertation to my wonderful father.

Richard Arthur Brinsley Perkins
1941-2009

A lover of life, an adventurer, a man who believed anything was possible.

So much of who I was, who I am now, and who I will be, both personally and professionally, is because of you.

Acknowledgments

I would like to convey my sincere gratitude to all my committee members:

Dr. William E. Haley: A master mentor! Thank you for always challenging me to exceed my own expectations. I am forever grateful to have been your student!

Dr. Matt Janicki: Thank you for always respecting my opinions, valuing my judgment, and supporting all my endeavors in the aging and intellectual disabilities field.

Dr. Sandy Reynolds: For all your enthusiasm, interest, and support over the years.

Dr. Brent Small: For your pragmatism, guidance, and encouragement.

Dr. Cathy McEvoy: Your office was always a safe haven during both good and bad times.

Dr. Krista Kutash: For being a wonderful chair at my defense.

Thank you to the faculty, staff, and students in the School of Aging Studies, and the Department of Aging and Mental Health Disparities, for your guidance, friendship, and support.

Finally, I'd also like to thank all the caregivers who participated in the study. You gave me such an enriching learning experience from your willingness to share the details of your lives. I am also indebted to all those that helped to promote my study.

Table of Contents

List of Tables

List of Figures

Caregivers of Adults with Intellectual Disabilities:
The Relationship of Compound Caregiving and Reciprocity to Quality of Life

Elizabeth A. Perkins

ABSTRACT

This study investigated the relationship between compound caregiving (i.e. multiple caregiving roles), and reciprocity to the wellbeing of older caregivers of adult children with intellectual disabilities. The study sample was composed of 91 caregivers with a mean age of 60 years. Participants were a convenience sample of caregivers predominantly residing in Florida. Care recipients' mean age was 29 years.

Thirty-four were currently compound caregivers. Quality of life indicators used as outcome measures in this dissertation were life satisfaction, depressive symptomatology, physical health, mental health, and desire for alternative residential placement of the care recipient. Compared with the non-compound caregivers, the compound caregivers had increased desire to place their care recipient into residential care. They also spent an average of 12 additional hours per week undertaking the compound caregiving role. Between group differences were not detected in life satisfaction, depressive symptomatology, global physical health, or mental health.

The role of reciprocity was investigated using tangible reciprocity (i.e. help with home chores), and emotional reciprocity (i.e. positive emotions). Overall findings indicated that caregivers reported giving more tangible and emotional support than they received, but considerable variability was evident. Relative disadvantage in tangible reciprocity was associated with increased depressive symptomatology, poorer mental health, and reduced desire for residential placement of the care recipient, but not with physical health or life satisfaction. Emotional reciprocity was not associated with any of the outcome measures.

Tangible reciprocity and compound caregiving were assessed using hierarchical regression analyses, to investigate their predictive value, after controlling for caregiver demographic variables, care recipient characteristics, and caregiving stressor variables, for mental health, depressive symptomatology, and desire for residential placement. Compound caregiving status was found to predict greater desire for placement over and above the control variables. Tangible reciprocity did not explain any significant variance in any of the regressions.

Overall, compound caregivers are more likely to desire residential placement for their care recipient, though no discernable difference existed between compound versus non-compound caregivers in the other outcome measures. Tangible reciprocity had little predictive utility in the present study. Compound caregiving research needs further refinement of more homogeneous groupings of compound caregivers.

Chapter One: Background

General Introduction

The central theme of this dissertation was to explore factors that may impact the quality of life outcomes for aging caregivers of co-residing adult children with intellectual disabilities (ID). In order to provide a comprehensive background of the extant literature, several areas need consideration. The background information describes terminology, historical perspectives, and demographic trends within the ID field. The next section provides an overview of general caregiving issues, including positive and negative physical and mental health outcomes noted in the broader caregiving literature, followed by ID specific caregiving issues. The Stress and Coping paradigm, is introduced as the guiding theoretical model for this dissertation, and its applicability to the design of caregiving intervention studies will be noted. This dissertation study addressed two important and unanswered questions in the field, namely the relationship of reciprocity and multiple simultaneous caregiving roles to the well-being of caregivers.

Overview of Intellectual/Developmental Disabilities

Developmental/Intellectual Disabilities Terminology

Within the field of developmental and intellectual disabilities there are a number of key terms that require definitions and clarification before proceeding any further.

The federal definition of developmental disabilities (DD) as defined by the

Developmental Disabilities Assistance and Bill of Rights Act of 1990 (PL98-527), refers to a severe, chronic disability of a person 5 years or older that

1. is attributable to a mental or physical impairment or combination of mental and physical impairments;
2. is manifested before the person attains age 22;
3. is likely to continue indefinitely;
4. results in substantial functional limitations in three or more of the following areas of major life activity: (a) self-care, (b) receptive and expressive language, (c) learning, (d) mobility, (e) self-direction, (f) capacity for independent living and (g) economic self-sufficiency;
5. reflects the person's need for a combination and sequence of special, interdisciplinary, or generic care, treatment, or other services that are lifelong or of extended duration and are individually planned and coordinated.

Generally, some common conditions that are considered developmental disabilities include cerebral palsy, epilepsy, autism, and intellectual disabilities. Approximately 2-3% of the population is comprised of individuals with DD (Crocker, 2006).

Though the terms developmental and intellectual disability are often used synonymously, there is a notable difference. Intellectual disability refers specifically to a subset of the population with developmental disabilities, whose major functional limitations is predicated primarily by intellectual and cognitive limitations. Thus, it is possible for a person to have a developmental disability, but not necessarily an intellectual disability, such as a person with cerebral palsy who has normal intellectual functioning. However, a person with ID always has a DD.

The American Psychological Association (APA) in the *Manual of Diagnosis and Professional Practice in Mental Retardation* defines mental retardation (i.e. ID), as (a) referring to significant limitations in general intellectual functioning; (b) significant limitations in adaptive functioning, which exist concurrently; and (c) onset of intellectual and adaptive limitations occurred before the age of 22 years (Jacobson & Mulick, 1996). Traditionally the level of intellectual disability present has been described in terms of mild, moderate, severe and profound (Jacobson & Mulick, 1996). These levels were primarily linked with the assessed IQ level where ID was apparently manifested, normally at 2 or more standard deviations (SD) of intellectual functioning from the population mean on standardized IQ tests. Also, the number of concurrent limitations in domains of adaptive functioning (AF) is also considered. There are three major domains of adaptive functioning; conceptual, social, and practical.

Mild ID has IQ scores within the ranges of 55-70 (<2 SD's, and 2 or more domains in AF). For moderate ID, IQ is within 35-54 (<3 SD's, and 2 or more domains in AF). For severe ID, IQ falls between 20-34 (<4 SD's, and all domains in AF). Finally, profound ID is defined by an IQ of 20 or below (<5 SD's, and all 3 domains of adaptive functioning), (Jacobson & Mulick, 1996). According to the *Diagnostic and Statistical Manual of Mental Disorders,* Fourth Edition, Text Revision (DSM-IV-TR; American Psychiatric Association (APA), 2000), in terms of the population of person's with ID, 85% have mild ID, 10% have moderate ID, 2-4% have severe, and 1-2% have profound ID. The true prevalence of intellectual disabilities has been difficult to determine, but is estimated in the DSM-IV-TR to be approximately 1% of the population. This equates to approximately 3 million people in the USA.

Historical Perspectives, Guiding Philosophies of Care, and the Impact of Deinstitutionalization

The residential status of persons with ID has varied considerably over time often reflecting the predominant guiding philosophy of care. Segregation in large scale institutions was once the norm due to overt medicalization of conditions associated with ID. This was a time when a generic plan of care was provided to persons with ID, and was not effective at recognizing individual differences. Persons with ID were routinely removed from their family home and community, with the encouragement of professionals, to enable the person with ID to be "properly" cared for in specialized medical settings. Unfortunately, most of these institutions were geographically isolated from their surrounding population leading them to become "closed" communities (Heaton-Ward, 1975). It is not unreasonable to assert that this segregation actually fostered further stigma and misunderstanding of people with ID.

Major changes in the last 30-40 years have been the result of a seismic shift in advocating for the fundamental and basic civil liberties that should be afforded to persons with ID. A major catalyst for this change was the United Nations Declaration of the Rights of Mentally Retarded Persons in 1971 as well as the philosophies of normalization (Wolfensberger, 1972), and "Social Role Valorization" in which "socially valued roles and life conditions for people" were demanded for all persons with ID (Wolfensberger, 1983, p.234). The major themes of the UN declaration and normalization philosophies, were that persons with ID have the same rights as all other human beings, have the right to education, training and therapy, have the right of economic security and access to meaningful occupation, and should live with their own families and participate in community life. These principles were also guided by the notion that as long as persons with ID were segregated from mainstream community, they would be perpetually stigmatized. Institutionalized living was not normal community living, and without living within the community, full participation and inclusion within the community would never be achieved.

Though these rights seem basic and self-evident, one must pause to consider that conditions in long-term institutions were, for the most part, deplorable and inhumane. Exposés of institutionalized life, both fictional (e.g. Kesey, 1962) and factual (e.g. Blatt & Kaplan, 1966) helped to publicize such atrocities where access to education, occupation, personal relationships, personal development and choice and autonomy over one's life direction was practically non-existent. Even basic choices for food, clothes, haircuts, leisure activities and adornment of one's living environment were either denied or seriously curtailed.

An important influence of normalization was that persons with ID are not sick per se, but do require lifelong care and support over their lifespan. The developmental model was presented as stark contrast to the medical model that viewed intellectual disability as an incurable medical issue (Wolfensberger, 1976). As described by Wolfensberger, the developmental model "…does not invest in the differentness of the retarded person with strong negative value. Even if severely retarded, he is perceived as capable of growth, development and learning", (1976, p. 44).

The result of the changes in philosophical approaches to the care and rights of persons with ID, led to the closure of many large institutions (i.e. deinstitutionalization) and extensive resettlement programs to community housing. Smaller more home-like community based settings with access to specialist medical/social services, when required, became widely advocated. The provision of community based residential homes has been implemented very successfully over the last few decades. Persons with ID now live in as many varied residential settings as the non-ID population. Many alternatives now exist, including independent supported living, small group homes, and staying within the family home. In fact, due to supports available to families, including access to special education and occupational opportunities, there has been a tremendous increase in the number of persons with ID who remain living at their parental home for their entire lifetime. Recent figures from 2006 suggest that of the 4.7 million persons with I/DD in the USA, 2.8 million were residing with family caregivers (Braddock, Hemp, & Rizzolo, 2008). Furthermore, 715,000 were residing with family caregivers aged 60+ (Braddock 1999; Fujiura, 1998; U.S. Census Bureau, 2007). Generally, the percentages for where persons with ID reside have remained fairly stable over the past decade. Around 60% live with a parent caregiver, 15% live with their spouse, 13% live independently, and 12% live in a supervised residential facility (Fujiura, 1998; Braddock,1999; Braddock, Hemp, & Rizzolo, 2008).

Among those who reside in residential facilities, around 20% live within "institutional" type residential facilities (state institutions and nursing homes that are serving 16 or more persons), and the majority reside in smaller group homes numbers (Braddock, Hemp, & Rizzolo, 2005). Between 1977-2005, the population within state institutions nationwide dropped 74%, an indicator of the success of resettlement policy nationwide (Lakin, Prouty, & Coucouvanis, 2006). Forensic considerations aside, it is a failure of current funding/support systems that there are still persons with ID who reside in an institutionalized setting. This happens when there is no state provided, agency provided or family provided residential alternative in their geographic locale to adequately serve their needs. Current institutionalized populations are usually persons with severe/profound ID with considerable psychiatric or physical comorbidities, and/or severe challenging behaviors. However, regardless of severity of ID and comorbidity, the vast majority of the ID population could be successfully supported in community settings, if there was sufficient care/funding mechanisms in place to meet their needs. Presently, it is also the case that alternative residential placements are subject to considerable waiting lists as demand currently outstrips current capacity due to restricted funding and lack of physical accommodation (Prouty, Smith, & Lakin, 2003).

Nevertheless, the residential status of persons with ID is one place where monumental change has been evidenced from guiding philosophies being embraced and implemented by policy and stakeholders. It is clear by far that living with one's parent is now the major residential status of the majority of the population with ID.

Demographic Trends

As with the general population, the life expectancy of persons with ID has risen substantially during the last century. Indeed, the increase in the ID population has been greater, and has occurred more recently and more rapidly. The development and implementation of holistic, individualized and proactive care philosophies has undoubtedly factored into the striking increase in longevity (Haley & Perkins, 2004). This is largely due to improvements in medical treatment of commonly associated comorbidities, such as cardiac problems and epilepsy (Haley & Perkins, 2004), as well as successful treatment of recurrent respiratory infections that was responsible for many premature deaths in childhood.

In terms of life expectancy, there is still disparity between the general population, persons with ID and persons with Down syndrome (Janicki, Dalton, Henderson, & Davidson, 1999), but the gain in the latter two populations has been substantial. To illustrate, between the 1930's and 1990's, the mean age at death for persons with ID increased 47 years, from 18.5 years to 66.2 years (Braddock, 1999). Persons with Down syndrome do still have a reduced life expectancy compared with others in the ID population but, nevertheless, they too have a much more extended lifespan. In the early 1900's life expectancy was merely 9 years (Selikowitz, 1990). From 1984 to 1993, mean age of death for those already 40 and above, had reached 55.8 years (Janicki, Dalton, Henderson, & Davidson, 1999). Due to genetic and endocrinological factors, people with Down syndrome (DS) are likely to retain their reduced life expectancy as their aging process is more precocious.

It is evident that the aging population of people with ID has risen at a greatly accelerated rate compared to the general population. Indeed, it has been suggested that for all other adults with ID, particularly those without serious associated physical/medical conditions, the disparity between their life expectancy and that of the general population will continue to decrease (Janicki, 1996).

Ramifications of Increased Longevity and Changes in Residential Status

Increased life expectancy and changes in residential status have transformed the lives of many persons with ID. Persons with ID are now a visible part of the community, and live predominantly with their families. Historically, it was the norm for many parents to survive their child with ID, but it is now apparent that more persons with ID are aging into older adulthood. Therefore this increases the likelihood of older adults with ID, who will outlive their parents.

The following crude comparison serves to illustrate this emerging trend. To argue this case effectively, one must also consider that increases in life expectancy (L.E.) has occurred in parents too. In both cases, the assumption is made that the parent has a child with ID, at age 25. A parent born in 1910 (L.E. = 51.5 years; Arias, 2006) who had a child in 1930's (L.E. = 18.5 years; Braddock, 1999) would themselves expect to live until 1961, but their child would have already died in 1953. Thus the average parent would have outlived their child by 8 years.

If we now consider a parent born in 1970 (L.E. = 70.8 years; Arias, 2006), who had a child in 1995 (L. E. = 66.2 years; Braddock, 1999), the parent would expect to die in 2040. However, their child would live to 2061. Therefore, an average child with ID born in the 1995 would now be expected to outlive their parent by 21 years. Thus it appears that many adult children with ID will remain living within their family home, receiving care and assistance from their aging parents, and increasingly face the prospect of ultimately outliving these exceptional caregivers. Research into caregiver quality of life is particularly significant, as more of these caregivers look set to devote the entirety of their life to this role.

Chapter Two: Overview of General Caregiving Issues

The Scope of Caregiving in the USA

The focus of this dissertation was the physical and mental well-being of family caregivers of adults with ID. It is useful to consider general findings about the caregiving experience as a precursor to the consideration of the caregiving experiences in the ID population.

The importance of caregiving cannot be more succinctly expressed than by the following quote by former first lady Rosalynn Carter, *"There are four kinds of people in this world: Those who have been caregivers, those who currently are caregivers, those who will be caregivers, and those who will need caregivers."* This quote hints at the pervasive nature of being a caregiver, knowing a caregiver, or needing a caregiver. Indeed, 44 million (over 1/5) of the US population are currently actively providing unpaid informal caregiving tasks to a relative or close friend age 18 and older (National Alliance for Caregiving & AARP, 2004). This figure does not include the substantial number of people who have previously undertaken caregiving responsibilities that have ceased, either due to transfer to long-term residential/nursing care, or from the death of the care recipient.

The care and responsibilities undertaken by caregivers on a daily basis is considerable, both in economic terms and its sociological impact. Perhaps the most important aspect of family caregiving is that it helps to maintain family structures, enables care recipients to remain in their home environment, and enjoy the benefits of receiving individualized attention (Perkins, Lynn, & Haley, 2007). After all, no one is more familiar with personal likes, dislikes, and mannerisms than their own family members who become their caregivers.

It has been estimated that if the informal unpaid services undertaken the family caregivers were provided by staff requiring remuneration, the cost would be a staggering 306 billion dollars annually (National Family Caregivers Association & Family Caregiver Alliance, 2006). Informal and unpaid family caregivers actually provide 78% of long-term care in the United States (Thompson, 2004). Compare this with those cared by combinations of formal and informal long-term care providers (14%) and formal providers (8%) alone (Thompson, 2004). Caregivers undoubtedly provide the mainstay of care, saving the economy billions of dollars, while maximizing the amount of time that care recipients can stay within their own home.

General Caregiving Research

Given the ubiquitous nature of the caregiving experience, considerable research has been undertaken to study the health impacts and quality of life issues that a caregiver encounters. Although caregiving research predominantly arose out of the study of caregivers to those with dementia, especially Alzheimer's disease (e.g. Zarit, Reever, & Bach-Peterson, 1980; Haley, Levine, Brown, & Bartolucci, 1987), it has rapidly broadened its sphere of inquiry to investigate the caregiving experience for a variety of illnesses and chronic conditions, such as as cancer (e.g. Vanderwerker, Laff, Kadan-Lottick, McColl, & Prigerson, 2005), Acquired Immune Deficiency Syndrome (Folkman, Chesney, Cooke, Boccellari, & Collette, 1994), stroke (McCullagh, Brigstocke, Donaldson, & Kalra, 2005), multiple sclerosis (Cheung & Hicking, 2004) and

Parkinson's disease (Martínez-Martín et al., 2005). Furthermore, unique subpopulations of caregivers to those with lifelong chronic impairment or disability have also received research interest. These include caregivers for adult children with severe mental health problems (McDonell, Short, Berry, & Dyck, 2003), and the subject of this dissertation, caregivers of adult children with intellectual disabilities (e.g. Chen, Ryan-Henry, Heller, & Chen, 2001; Haley & Perkins, 2004).

Physical, Psychological, Social, and Financial Consequences of Caregiving

Even though research has generally broadened to look at the issues that caregivers encounter from particular conditions or subpopulations discussed above, one is able to identify four common domains that have been noted to lead to adverse consequences across most caregiving situations. They are the physical impacts of caregiving, the psychological impacts of caregiving, the consequences of caregiving on one's own family and social life, as well as the financial implications of performing a caregiving role long-term. These domains can transpire irrespective of the particular idiosyncrasies of each caregiving circumstance.

Regarding physical health, earlier studies were quick to identify that the considerable stress of caregiving can lead to adverse impacts on one's physical wellbeing in clinically measurable ways. Caregiving stress can result in blood pressure elevations (King, Oka, & Young, 1994) and increased insulin levels (Vitaliano, Scanlan, Krenz, Schwartz, & Marcovina, 1996), and greater risk of developing cardiovascular disease (Lee, Colditz, Berkman, & Kawachi, 2003). Caregiving stress can lead to compromised functioning of the immune system (Kiecolt-Glaser, Dura, Speicher, Trask, & Glaser, 1991), and more specifically, increase the healing time for standardized wounds (Kiecolt-

Glaser, Marucha, Malarkey, Mercado, & Glaser, 1995). The stress of being a caregiver can therefore reduce the ability of one's own immune system to function optimally, making the caregiver more susceptible to infections and illness. One of the most disturbing findings in this line of research was that merely being a highly strained caregiver was found to be an independent risk factor for significantly elevated mortality over several years (Schulz & Beach, 1999).

The potential adverse consequences of caregiving on psychological health have also been investigated. Caregivers have been noted to have an increased risk of developing depressive disorders (approximately 30%) or suffering from significant depressive symptoms (55%), when they are compared matched control groups and population norms (Schulz, O'Brien, Bookwala, & Fleissner, 1995; Haley & Bailey, 1999). Caregivers also report significant decrements in other indicators of psychological well-being. Compared with noncaregiving controls, caregivers report increased feelings of stress, lower levels of self-efficacy and lower levels of subjective well-being (Pinquart & Sörensen, 2003).

Physical and psychological impacts aside, caregiving has also been acknowledged to have very serious implications for the quality of relationships and cohesiveness between the caregiver and other close members of the caregiver's family. One notable consequence is that performing caregiving duties can drastically reduce the amount of time available for the caregiver to interact with their other family members, and friends (NAC & AARP, 2004). Supervision of a care recipient often leads to substantial reductions in the vacations, hobbies, and leisure activities that a caregiver can independently undertake, a consequence that may worsen over time as the care recipient becomes increasingly incapacitated (NAC & AARP, 2004). Furthermore, as social contacts become increasingly difficult to be adequately nurtured, and social interaction and participation in social activities declines, caregivers are, unsurprisingly, at greater risk for increasing social isolation, often resulting in a substantial reduction in social support over time (Haley & Bailey, 1999; Robinson-Whelen, Tada, MacCallum, McGuire, & Kiecolt-Glaser, 2001). The fact that the time when caregiving duties are getting more demanding, is likely to coincide with less social support being readily accessible, is one of the most unfortunate social outcomes of the caregiving experience (Perkins, Lynn & Haley, 2007).

Finally, the decision to undertake caregiving duties can result in substantial financial penalties to the primary caregiver and their family (Langa et al., 2001). Financial difficulties can arise from being unable to fully participate in the workforce, due to inflexible scheduling demands and the inability to undertake work outside of the home setting. Another factor is that the level of supervision and care provided depending upon the caregiving scenario can be unpredictable, again leading to difficulty in securing continuous employment with benefits, or cessation of employment due to increased caregiving responsibilities (e.g., Schulz et al., 2003). A further consideration is that an alternative caregiver is required to fill-in for the primary caregiver. If a formal caregiver is employed, then the financial benefit derived from the outside employment is reduced. Otherwise, one has to find an alternative caregiver, who is someone who also has the time and resources to freely commit to the caregiving duties in the absence of the primary caregiver.

The "career" of a caregiver can span many years and therefore the earning capacity of a primary caregiver, coupled with out of pocket expenses that can accrue with caregiving, can greatly interfere with the financial status of the primary caregiver and indeed the rest of their family. This can place a chronic financial strain on the caregiver, which can lead to an inability to invest for their current personal and future retirement needs, or even the abandonment of planned retirement in response to the caregiving role.

Positive Experiences of Caregiving

Even though the foregoing could be stated to depict the worse case scenarios for adverse outcomes for caregiving, it should also be acknowledged that being a caregiver can and does bring intrinsic rewards. The role of caregiver can be very empowering to one's sense of self esteem (Nijboer, Triemstra, Tempelaar, Sanderman, & van den Bos, 1999). Caregivers report that the role of caregiver makes them feel useful and needed, increases their appreciation of life in general, helps them actually develop a more positive attitude, and often helps strengthen the bonds of their relationships (Tarlow et al., 2004).

Family Caregiving for Adults with Intellectual Disabilities

Potentially Adverse Factors of Family Caregiving, Aging, and ID

Although aging family caregivers of adults with ID are likely to have similar challenges as those in the general caregiving community, nevertheless, they are also a unique group of the caregiving population that also face distinctively different circumstances from caregivers of persons that have developed illnesses, or disabilities that have arisen from accident or trauma. The extended duration of the caregiving role, additional health issues of the aging care recipient, and anxiety for the future well-being of the care recipient, are three factors that need to be considered (Haley & Perkins, 2004).

Extensive duration of the caregiving role.

The average caregiving career, i.e. the period of time that a caregiver provides assistance to a spouse or older family member with a chronic illness is approximately 4.3 years (National Alliance for Caregiving & AARP, 2004). However, caregiving for a child with ID is, unless interrupted by alternative residential placement, a lifelong endeavor which spans many decades. Indeed, these caregivers may never have any period in their life free from caregiving responsibilities and roles. One can say that a parent is a parent for their lifetime, but in the vast majority of cases, at some point children become financially independent, co-residence ceases, and children will forge their own direction in life and start their own families, independent of their parents. For the majority of persons with ID, residence with family caregivers will continue until either alternative residential placement is sought, either for upholding a sense of independence for the adult with ID, or when the caregiver is no longer willing or able to perform their role due to illness or disability in either or both caregiver and care recipient.

Caregiving for people with ID may result in increased vulnerability and social isolation purely because of the extended period that caregiving is undertaken. Seltzer, Greenberg, Floyd, Pettee and Hong (2001) analyzed data from the Wisconsin Longitudinal Study (WLS) of parents who had a child with ID compared to non-ID control parents. The WLS has tracked 10,000 adults, and interviewed them at age 18, 36, and 53-54 years. Seltzer et al. found that parents of children with ID reported significantly lower rates of social participation compared with control parents (2001).

Todd and Shearn (1996) recognized the importance of cognitive appraisals expressed by parents of children with ID regarding their caregiving role, an important consideration given its potential lifelong duration. Todd and Shearn categorized parents as being either "captive" or "captivated". Captivated caregivers endorse the fact that they have embraced their role, and found much contentment and purpose because of it. On the other hand, captive caregivers feel trapped in their role, and that their own life ambitions have been overshadowed. These categories were utilized in a study of the quality of life of ID parental caregivers and it was reported that captive parents reported significantly higher levels of parental stress, and pessimism (Walden, Pistrang, & Joyce, 2000). One fundamental difference noted between the captive and captivated parents was that the children of captive parents exhibited significantly more challenging behaviors. Challenging behavior has long been noted in previous research to be associated increased stress and burden in caregivers of persons with ID (Black, Cohn, Smull, & Crites, 1985; Dumas, Wolf, Fisman, & Culligan, 1991; Grant & McGrath, 1990; Heller & Factor, 1993). In the ID field, challenging behaviors refer to a set of problematic behaviors that place both the caregiver and their care recipient at risk for adverse outcomes for their health, safety and welfare (Emerson, 2001), as well as often resulting in social isolation. In particular, challenging behaviors include aggression, destructiveness, overactivity, self-injury, inappropriate sexual contact/conduct, bizarre mannerisms, and eating inappropriate objects/substances (i.e. pica) (Emerson, 2001).

Additional health concerns due to aging process in care recipient.

Over the course of their child's lifetime, a caregiver may have already faced an array of medical crises and might be well adapted to coping with changes in associated physical disabilities and health conditions that have arisen over the years. However, physical changes due to aging, or aging-related health conditions can present the ID caregiver with a number of additional challenges. As Evenhuis, Henderson, Beange, Lennox, and Chicoine (2001) noted, greater longevity can also bring additional functional impairment, morbidity, and mortality from early age-onset conditions, from both their progression over the lifespan, and also their interactions with older age-onset issues.

For example, sensory impairments are particularly problematic to the aging ID population (Wilson & Haire 1990; Warburg 1994; Schrojenstein Lantman-de Valk et al. 1997). Generally, aging-related changes in vision (presbyopia) and hearing (presbycusis) may be present in rates similar to those in the general population, as are age-related pathologies of vision (e.g. cataracts, macular degeneration, glaucoma, and diabetic retinopathy), but the impact is often more severe due to higher rates of pre-existing, childhood onset of visual and auditory pathology (Schrojenstein Lantman-de Valk et al. 1994; Evenhuis, 1995a, b).

Musculoskeletal issues also arise more frequently. By the age of 60, around 30% of people with ID will have significant mobility and gait issues, and by 75 years, 60% will (Evenhuis, 1999). Issues such as endocrinological dysregulation and hypotonia in persons with DS, and musculoskeletal deformities and limitations that restrict weight bearing exercise in persons with cerebral palsy results in increased prevalence of osteoporosis, and furthermore the overall ID population also has higher risk of osteoporosis compared with the general population (Center, Beange, & McElduff, 1998). Furthermore, although musculoskeletal abnormalities may not cause pain during child and early adulthood, persons with cerebral palsy are much more likely to develop osteoarthritis at a younger age. This is due to restricted patterns of movement, as well as abnormal contact and compression between joint surfaces.

Adults with DS are at a significantly greater risk of developing Alzheimer's disease. One statewide study reported a prevalence of 56% of adults with Down syndrome over age 60 had dementia, compared to only 6% of all other adults with ID (from various etiologies) over age 60 (Janicki & Dalton, 2000). The latter group having prevalence rates the same as those reported for the general population (World Health Organization, 2001).

Various treatment modalities that are utilized over the lifespan can also have long-term consequences (Evenhuis, Henderson, Beange, Lennox, & Chicoine (2001). Bone mineralization problems can develop from chronic use of some anticonvulsants (Phillips, 1998), and particularly troublesome tardive dykinesias can arise from long-term neuroleptic use (Haag et al., 1992; Wojcieszek, 1998). The foregoing is just a small sample of many aging issues that can arise with this population.

A cumulative effect of long-term caregiving may also result in an increased risk of adverse health outcomes in the caregiver due to the physical performance of some caregiving tasks. Those caring for people who are immobile, or are in other ways highly dependent in achieving the basic activities of daily living, may be especially susceptible. Indeed, there has been recent controversy regarding a prepubescent girl with ID, who is being treated with growth attenuation drugs, to prevent her from reaching biological puberty, as well as having a hysterectomy. Her parents were fully supportive of her treatment claiming that stopping further growth would make their future lives as long-term caregivers more manageable, and reduce the risk of out-of-home placement. The ethical debate this ignited led to a position statement against the practice being released by the American Association on Intellectual and Developmental Disabilities (2007). The debate still continues, but the "growth attenuation treatment" is a direct result of the recognition of potential hardship caregiver's face in being physically able to maintain caregiving duties over time that prompted this course of treatment.

The care recipient is not the only person that has to deal with the onset of aging-related health conditions. Aging caregiver themselves can also be affected by the increased incidence of aging-related chronic illnesses of later life. Increasing medical demands of the care recipient can also coincide with similar issues occurring in the caregiver. This can result in a reduction in the caregivers' physical ability to perform caregiving tasks, at a time when the care recipient is becoming more dependent.

Fears about the future long-term care of the care recipient.

Caregivers of advancing age at some point needs to address two major questions regarding their future needs of their child, especially in the event of their own deaths. Two major questions facing aging parents are "Where and with whom will my child live after I am gone?", and "Will my child have adequate financial security?" Grant (1990) noted that a common reaction is for parents to essentially freeze up with indecision when considering the future, especially when viable solutions do not appear to be readily available due to prohibitive costs, long waiting lists, or unsuitable residential alternatives. One study that specifically asked aging mothers (aged 58-87) whether they had made the necessary residential plans for their children, found that less than 50% had actually done so (Freedman, Krauss & Seltzer, 1997). It appears that lack of future planning is quite widespread.

Essex, Seltzer and Krauss (1999) reported that fathers of adults with ID are much more pessimistic than their wives about their child's future care. It was noted that such pessimism may be the result of considering themselves to have primary responsibility for providing the financial means for future care provision. Kelly and Kropf (1995) noted two aspects that can impact the financial stability of a family in these circumstances. Firstly, caregiving can result in a substantial loss to overall household income, due to caregiving duties impacting earning potential. Secondly, the loss of earning capacity may have drastically impacted the ability for saving and investments for the future retirement needs of the whole family. For example, using data over a 35 year period from the WLS, it was found that both parents, especially mothers, had significantly lower rates of employment, and when they were employed, experienced difficulties due to conflicting commitments to family and work (Seltzer et al., 2001). A later study using the same data set looked at income differentials over the same period (Parish, Seltzer, Greenberg, & Floyd, 2004). It was found that when compared to non-ID parents, ID parents' mean annual household income by the time they were 53 years old, was 31% lower. In terms of savings, parents of children with ID had mean savings 36% lower ($132,700) than the $181,000 for control parents.

Not only does ID caregiving have an impact on the earning capacity of the family, it also incurs substantial additional economic expenditures to the family in terms of transportation, medical costs, and other specialized services. More specifically, these include respite care, specialized therapies, home modifications and adaptive technologies, medications and occupational/educational services (Parish, Seltzer, Greenberg, & Floyd, 2004). One study estimated such "out of pocket" expenses to average 16.9% out of total household income (Fujiura, Roccoforte & Braddock, 1994). More recently, it has been found that even middle-income families of children with ID, suffer considerable financial hardship (Parish, Rose, Grinstein-Weiss, Richman, & Andrews, 2008). Although persons with ID who meet strict disability and income criteria are able to receive Supplemental Security Income (SSI) payments, it is widely recognized that SSI is usually inadequate to cover the expenses accrued (General Accounting Office, 1999).

The foregoing highlights that caregivers of persons with ID can have substantial reduction in their earning capacity, when at the same time, the child with ID presents additional financial costs that are not adequately met by government and state-provided funding sources. Chronic financial strain over substantial periods of time is a stark reality for many of these caregivers. As these caregivers age, their financial circumstances may result in delaying their own retirement. Thus, fear of the future due to uncertainty of future residential opportunities, as well as financial insecurities can weigh heavily, particularly upon aging caregivers for whom these particular issues are increasingly imminent.

Potentially Beneficial Factors of Family Caregiving with Aging and ID

The foregoing can paint a very pessimistic picture but it is also the case that caregivers do derive personal benefits. Some researchers have long asserted that there has been too much importance placed upon caregiving as a pathological issue while ignoring the gratification that can be gained from the experience (Grant, Ramcharan, McGrath, Nolan, & Keady, 1998). Though there is just cause for concern from the issues that challenge caregivers of persons with ID, there are several mitigating factors that may positively impact the ID caregiving experience. The normative nature of caring for ones own child and expertise gained from lifelong caregiving are most notable (Haley & Perkins, 2004).

Normative nature of parental caregiving.

Caregivers of parents or spouses, especially those with dementia, are thrust into roles that for many would certainly require some period of adjustment for both caregiver and care recipient. Obviously, the quality of the relationship prior to undertaking caregiving responsibilities can greatly impact the caregiving experience. There are often fundamental changes in the dynamics of the relationship that evolve with increasing dependency and loss of autonomy of the care recipient. Some caregivers find it easy to provide care in the earlier stages, until the care recipient becomes increasingly incapacitated. These caregivers may be good supervisors and errand runners, but are very uneasy with the thought of having to provide more intimate and personal care needs, such as washing, bathing, or toileting the care recipient, or changing incontinence pads.

The spousal caregiver, or the adult child caregiver, in many respects take on a "parenting" type of persona. They become increasingly responsible for arranging and providing all aspects of care, as well as maintaining the home environment. Conversely, parents caring for an adult child find themselves in a more normative and familiar territory than those "new" caregivers for a parent or spouse. There are no new fundamental role dynamics at play, merely continuation of a relationship dynamic that has existed and developed over decades. Though many caregivers of persons with ID might also face changing medical needs of the care recipient, they are also the same caregivers that have already had a lifetime of providing highly specialized personal care.

Expertise and feelings of mastery gained from long term caregiving.

When a caregiving career has spanned several decades, there has no doubt been many challenges that have been faced and dealt with. Indeed, many caregivers gain extensive knowledge and expertise from facing these challenges and familiarity of their circumstances and the needs of their care recipient. There is no doubt that even for caregivers that have difficulties, they have successfully adapted and adjusted to their roles over time (Townsend, Noelker, Deimling & Bass, 1989). It is possible that a sense of mastery through experience, bestows an advantage on older caregivers when dealing with fresh challenges. Heller, Rowitz, and Farber (1992) found caregivers of adults aged 30 years and older reported significantly less burden than caregivers of those younger. Hayden and Heller (1997) also found that older caregivers report less burden, despite there being no significant differences in support services received, or the size of their social networks.

The foregoing background information has provided a broad picture of the general factors which can impact ID caregiving experiences. Many of the stressors that affect caregiver well-being tend to be unmodifiable, i.e. care recipient characteristics such as challenging behavior, level of support/assistance required, secondary health conditions that commonly arise in those with ID, early-onset aging-related illnesses, and caregiver's financial status. There is also the issue that in most caregiving scenarios, it is a time-limited endeavor, whereas in ID caregiving, these stressors are not only unmodifiable, but are of prolonged duration.

However, the difficulty in modifying certain stressors can be offset by applying the stress-process model to caregiver interventions aimed at improving caregiver well-being. The general philosophy of the stress-process model is to change one's perception of a stressor, and the resources available to cope with stressors, rather than approaching the issue only by efforts at reducing or eliminating the stressor. This is particularly effective for caregiving stressors that are not in themselves amenable to change.

Application of the Stress and Coping Paradigm to Study Caregiving

The outcomes of caregivers have long been studied under the theoretical guidance of Lazarus and Folkman's (1984) model of stress and coping (e.g. Haley, Levine, Brown, & Bartolucci, 1987; Harwood, Ownby, Burnett, Barker, & Duara, 2000; Pearlin, Mullan, Semple, & Skaff, 1990). Lazarus and Folkman defined stress as "a particular relationship between the person and the environment that is appraised by the person as taxing or exceeding his or her resources and endangering his or her well-being" (p. 19). The perception that an individual may experience in response to a particular situation/event (i.e. potential stressors, e.g. caregiving duties) is classified under the model as being irrelevant, benign, or stressful. If a caregiver perceives some aspect of their role as stressful, they make a subjective judgment or appraisal. Depending upon the circumstances, an individual can appraise the situation/event as either challenging, harmful, or threatening. Subjective appraisals ultimately guide the individual's response to stress, therefore this explains diverse reactions from people when faced with the same set of stressful circumstances.

The individual appraisal of a stressful situation describes how an individual views a potential stressor. However, the real utility of the stress-coping model is that it describes coping strategies that can be utilized in order to confront the potential stressor. According to Folkman and Lazarus (1980) there are two major types of coping, emotion and problem focused coping. The function of emotion-focused coping is to regulate distressing emotions, an example of which is finding empathy in a support group. Problem-focused coping is when direct action is undertaken to change the distressful situation (e.g. utilizing behavioral techniques to address challenging behaviors of the care recipient).

Chronic strains and stress that can accompany caregiving may not always be eased by using emotion-focused or problem-focused coping. In such cases, psychological distress can ensue, and stress can proliferate over time. Despite seemingly intractable situations, it was recognized that some caregivers are very resilient, even when it seems that neither emotion-focused nor problem-focused coping strategies are effective. This finding prompted the addition of meaning-based coping to the stress and coping model (Folkman, 1997). Examples of meaning-based coping include using one's spiritual beliefs, reevaluating the caregiving experience as a period of great personal growth, and revising one's goals to regain purpose and control. Lazarus and Folkman's (1984) model essentially emphasizes the importance of individual appraisal of stress rather than the occurrence or severity of the potential stressor per se.

Lazarus and Folkman (1984) categorized coping resources as being either internal (e.g., use of benign appraisals, sense of meaning, problem solving skills, previous experience, personality) or external (e.g. appropriately modifying an environment, social support, money). Internal resources are inherent within the person themselves, whereas external resources are accessed by changing the environment or utilizing assistance from others.

Caregiver stress and coping can be considered to be a delicate balancing act (see Figure 1), in which high levels of stressors can increase the likelihood of negative caregiver outcomes, while high levels of resources can help to decrease the likelihood of caregiver distress (Perkins, Lynn, & Haley, 2007). Pearlin, Mullan, Semple, and Skaff (1990) view caregiving as being an activity that has primary and secondary stressors. They describe the day to day care tasks (e.g. managing challenging behavior, bathing, feeding, dressing the care recipient, etc) as primary stressors. Secondary stressors are spillover effects from the primary caregiving role, and include marital disharmony, stress in other family relationships, and issues with occupational roles. Caregivers are also not immune from other life stressors and strains outside of their primary caregiving role (e.g. bereavement, house relocation).

Interventions for caregivers can enhance caregiver well-being by minimizing stressors (e.g. implementing a behavioral program to minimize challenging behavior), improving internal resources (e.g. altering appraisals, teaching alternative coping skills), or enhancing external resources (e.g. modifying the physical environment, increasing the level of social support). When assessing stressors that can impact caregiver well-being, it is important to consider not only primary and secondary, and other life stressors, but how caregivers actually appraise these stressors, and the internal and external coping resources that they utilize.

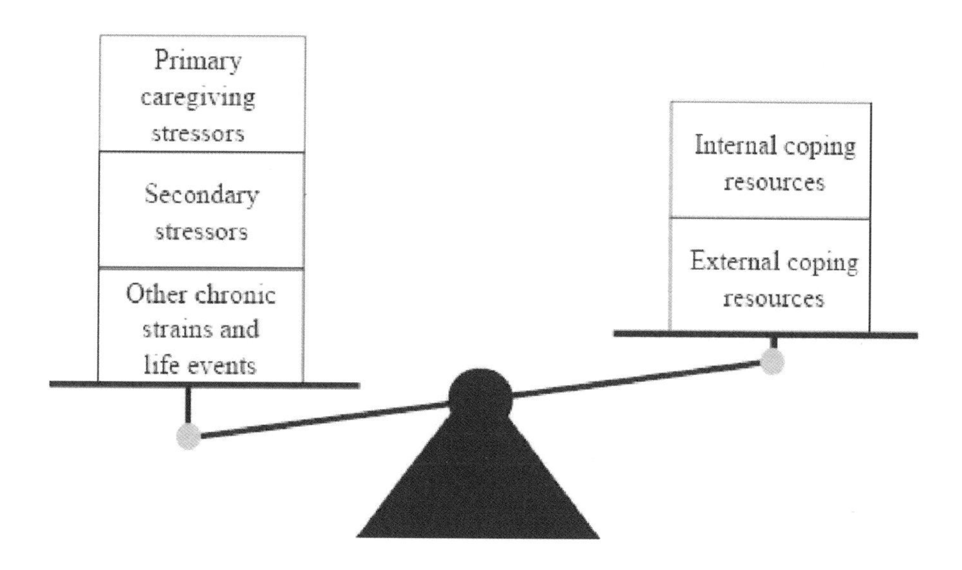

Figure 1.1. Stress and Coping: A Balancing Act (from Perkins, Lynn, & Haley, 2007).

Caregiving Interventions

The ultimate goal of caregiving research is to provide evidence-based interventions to alleviate the stress and burden of caregivers and improve quality of life outcomes, though there is often debate on what type of intervention is most beneficial. Sörensen, Pinquart, and Duberstein (2002) conducted a meta-analysis of the efficacy of 78 caregiver intervention studies across many caregiving scenarios to investigate this issue. Overall, the best results were found when using individualized and well-structured interventions. In particular, psychotherapy and psychoeducational interventions were the two most effective types, with positive improvements occurring in five clinically relevant domains of caregiver well-being and patient functioning. These were ability/knowledge, well-being, depression, burden, as well as improvement in care recipients' symptoms. Cognitive-Behavioral Therapy was the most common type of psychotherapy used in intervention studies, and included counseling on challenging negative thinking and assumptions, increasing pleasant activities, and teaching effective time management skills. Psychoeducational interventions were undertaken in group settings, and disseminated information regarding a care recipient's disease process, as well as providing training and resources to manage particular issues (e.g. challenging behaviors). The information was presented in structured formats including formal lectures, supplied written materials as well as group discussion. Education was the major component of both psychotherapy and psycho-educational therapy. In psychotherapy, the focus of intervention was directed towards increasing a caregiver's self-knowledge. In the psychoeducational approach, the aim is to increase the knowledge of the care recipient's disease and symptoms.

The stress-coping model incorporates many of the attributes that are found to improve caregiver well-being in interventions. For example, psychoeducational interventions are an example of problem-focused coping that increase internal resources (i.e. by providing strategies to deal with different aspects of caregiving roles). Psychoeducational interventions also improve access/knowledge to external resources that caregivers may benefit from. Cognitive-behavioral therapies work by modifying the cognitions (i.e. appraisals) of caregivers, and thus enhancing internal resources. Stress and coping models emphasize that the ability to successfully combat caregiving stressors is a function of enhancing the coping resources of the caregiver, rather than the apparent vulnerability of a caregiver (Vitaliano, Zhang, & Scanlan, 2003). This is an important aspect of intervention efficacy, because the ability to improve caregiver well-being depends more on improving coping resources, rather than minimizing vulnerabilities of caregivers that may actually differ dramatically within each intervention group.

As the literature on caregiving research has provided numerous examples of effective caregiving interventions that are theoretically guided on the stress-process model, this dissertation investigated two aspects of ID caregiving that have received little research attention. In reviewing the literature, there are several gaps in the knowledge of ID caregiving that have yet to be adequately investigated. However, given the utility of the stress-process model to provide a sound theoretical basis for effective caregiver interventions, it would appear that variables that can be investigated under this model would ultimately provide potential effective targets for future intervention studies. As such, this dissertation focused on both sides of the caregiving balancing act (see Figure 1). It assessed the impact of a chronic strain/life event that older parent caregivers may be particularly more exposed to, i.e. the fact that aging ID caregivers are sometimes having to undertake multiple caregiving roles for their own parents. This is an example of life event that may adversely magnify caregiving stress. There had been little investigation of the function of reciprocity (i.e. the feeling of a mutually beneficial exchange that occurs in all relationships). Reciprocity may potentially be regarded as a stressor (if lacking) or a coping mechanism.

Impact of Compounded Caregiving Duties on Aging Caregiver

An aspect that could adversely affect well-being in older caregivers is their experiences as sandwich caregivers. Sandwich caregiving refers to those people who undertake caregiving responsibilities for older adults while still having parenting duties to their own (usually teenage or older) children. Some of the major factors that make simultaneous parenting/caregiving an issue is the financial drain of a child, errand running, and maintaining responsibilities for non-coresiding children (due to divorce or school attendance) (Loomis & Booth, 1995).

These circumstances may actually not be too taxing in many cases, and, indeed, there is considerable controversy between researchers as to whether the sandwich caregiving generation is really as "beleaguered" as previously thought (Loomis & Booth, 1995). Conversely, there has been more recent speculation that "sandwich caregiving" does indeed exist. Furthermore, it appears that the period of time when someone can be a sandwich caregiver is increasing, due to the combined effect of baby boomers having delayed child-rearing and their own parents' increased life expectancy (Rogerson & Kim, 2005). Rogerson and Kim state that baby boomers are actually having a "stretched period of caregiving" compared with their own parents, and the same factors (delayed child-rearing and increased life expectancies) will also result in the children of baby boomers being sandwich caregivers for even longer.

Though there remains argument regarding the relative impact of sandwich caregiving overall, one scenario that has not been adequately explored is unique to ID caregivers. "Compound caregivers" are those parents who are already providing considerable caregiving responsibilities for their own child with ID, who subsequently becomes a primary caregiver for another family member (Perkins, in press).

What makes these compound caregivers significantly different from traditional "sandwich caregivers" is that they are already caring for highly dependent children, when the additional caregiving role is adopted. Furthermore, a recent case study has highlighted that periods of compound caregiving may occur several times, and may include caring for older parents, parent-in-laws, and siblings, in some cases (Perkins, in press). The same case study also reports that such periods can be extremely stressful, partly because many difficult decisions about prioritizing caregiving tasks had to be made.

Previous studies have, understandably, focused on the caregiving experience and tasks in terms of dyadic relationships. There has been little attention paid to assess whether the primary caregiver had responsibility for other care recipients either currently, or in the past, or indeed, whether they are anticipating a likelihood of additional care recipients in the future. This previous omission, though understandable, deserves further exploration especially as compound caregiving is likely to be an increasingly common scenario for ID caregivers given the increased life expectancy of people with ID, and the general population. Research is required to identify compound caregivers, to determine the frequency of compound caregiving, and quality of life outcomes for compound caregivers, during the compound caregiving episode, and long-term ramifications of these highly demanding periods.

Investigation into the Function of Reciprocity

Social exchange and equity theories explain why and how relationships are formed and maintained. At the most pragmatic level, it emphasizes that exchanges or relationships are formed between two parties in which both are seeking to maximize the benefits, while minimizing the costs (Adams, 1965; Cook, 1987; Molm & Cook, 1995; Walster, Walster, & Berscheid, 1978).

Equity theory is more concerned with the affective component of social exchanges. Therefore, it proposes that feelings of well-being are dependent upon individuals feeling equitably treated (i.e. they give and receive in proportional amounts) (Walster, Walster, & Berscheid, 1978). If balance is not achieved then it can lead to feeling of distress, but can happen when people feel that they are receiving too much (over-benefit) as well as too little (under-benefit) from the relationship (Walster et al.). If people feel they are receiving too much, they may feel guilty, whereas a person who feels they are not receiving enough, may become angry or resentful. Basically, there is the feeling that fairness should permeate relationships, and that adjustments are made to rectify perceived imbalances. Equity theory can be applied to all types of relationships, from occupational, employer/employee, to more intimate social relationships.

Most social relationships operate under the norms of reciprocity, in that there is a mutual exchange of emotions/services that is beneficial to both parties, which helps to initiate, but more importantly, fosters the maintenance of social ties (Levi-Strauss, 1964). These reciprocal exchanges occur both in the context of normal familial intergenerational assistance, and family caregiving (Walker & Pratt, 1991). However, Gouldner (1973) asserted that reciprocity is not unconditional, and that the status of those engaged in a reciprocal relationship can affect the function of the reciprocity. Gouldner thus argued that reciprocity may not apply to certain groups of people, the elderly frail being a particular population he cites. In the case of older adults, he suggests that the norm of beneficence is at play, whereby assistance is offered to help those in need, and that recipients offer gratitude to their caregivers but no reciprocity is evident.

However, as previously reviewed earlier, caregivers do derive personal benefits from their caregiving role. Thus, even if the norm of beneficence is initially responsible for initiating caregiving duties, reciprocity in caregiving may override beneficence as the mechanism which actually maintains the caregiving function. Therefore it can be argued that many of those who initially undertake caregiving duties do so because they feel compelled to help a relative who needs assistance. However, not all caregivers are able to cope and continue with their role, so despite their best intentions, beneficence alone, does not explain how someone becomes a successful caregiver. Indeed, reciprocity may be a mechanism that contributes to both successful and sustained caregiving.

Previous studies of reciprocity.

The positive association between reciprocity and well-being has been found in several studies regardless of the type of relationship, including occupational relationships (e.g. Bunk, Doosje, Jans, & Hopstken, 1993), and close friendships (e.g. Buunk & Prins, 1998; Rook, 1987). In a study that tested whether reciprocity or social support that was given or social support received reduced symptoms of stress, it was found that reciprocity from family members was a significant predictor, but not the social support measures (Jung, 1990). Reciprocity was also noted to be an important factor in explaining within-family variation in support that mothers provided differentially to their own adult children (Suitor, Pillemer, & Sechrist, 2006).

In marital relationships there have been conflicting results. One study did find a positive association of reciprocity to well-being (e.g. Buunk & Mutsaers, 1999), whereas balanced reciprocity within marriage had no association to the improvement of general health (Väänänen, Buunk, Kivimäki, Pentti, & Vahtera, 2005). However, a large scale epidemiological study that investigated the importance of reciprocity to one's overall health (Kawachi, Kennedy, & Glass, 1999), did find that people who trusted that those closest to them would engage in the norms of reciprocity, were 1.7 times more likely to have good or very good self-rated health, compared with those who did not believe this would be the case.

Within the caregiving context, early studies have also indicated that the level of reciprocity a caregiver perceives is an important consideration in their well-being. For example Hirschfield (1983), found that when a greater sense of mutuality and reciprocity was expressed by caregivers, it was associated with increased sense of coping with the responsibilities of their role, and decreased their desire to consider institutionalizing their care recipient. Neufeld and Harrison (1995), in their study of dementia caregivers, found that the caregivers actually monitored the give and take in their relationship to the care recipient and that many of them "developed a constructed reciprocity that was built through monitoring and included the subprocess of observation" (p. 355). The caregivers who were able to construct such reciprocity with the care recipient were more satisfied and content with themselves, i.e. experiencing a sense of reciprocity actually helped to bolster their self esteem. Carruth, Tate, Moffett, and Hill (1997), reported that reciprocity accounted for significant variance in family satisfaction reported by caregivers of elderly parents. A more recent study, by Reid, Moss and Hyman (2005) also found that those with higher levels of reciprocity reported reduced caregiver burden.

Similarly, as one would expect, if one considers that equity theory suggests that reciprocity helps to maintain balance, a few studies have taken into account care recipient's experiences. Care recipients have reported significantly less depressive symptomatology when they have reported greater reciprocity (Pruchno, Burant, & Peters, 1997; Wolff & Agree, 2004). According to Gouldner (1973), frail care recipients could do little but offer gratitude, not reciprocity, but would offering gratitude alone explain reduction in depressive symptoms? If care recipients were only able to offer gratitude, caregivers may still perceive imbalance within the relationship, and such imbalances are thought to lead to greater distress. Though many of the previous studies discussed seem to indicate that more emotional types of reciprocities have been associated with improved sense of well-being, there are also reports of caregivers experiencing lower levels of stress and burden, when their care recipient is able to provide actual tangible help with various chores (Dwyer & Miller, 1990; Dwyer, Lee, & Jankowski, 1994). This would also seem to contradict Gouldner. The actual amount of tangible reciprocity received may very well be minimal, token even, but nevertheless in the caregiver's eyes it is seen as significant. Even the most minor task may have significant benefit on the feeling of reciprocity that a caregiver expresses. Perhaps the mere act of the care recipient trying to help reduces the caregiver's appraisal of their dependency level.

As far as studies of reciprocity for caregivers of persons with ID, it is an aspect that has only been briefly acknowledged. Heller, Miller, and Factor (1997), did investigate "support from adult child with mental retardation" as an independent variable to predict perceived caregiving satisfaction. Even after controlling for other forms of informal support, caregiver/care recipient health variables, and challenging behaviors, support from the adult child still accounted for significant variance in caregiving satisfaction. Heller et al. noted that the adults with ID provided their parents with an enduring source of companionship and constructive help with household tasks.

Though there is certainly not an extensive literature on reciprocity, there appears to be mounting evidence of the benefits of assessing reciprocity given that it is associated with positive outcomes for caregivers. Furthermore, although there is far less research that has studied care recipient well-being, the indication is that high levels of reciprocity are also associated with better outcomes.

Typologies of reciprocity.

Although the above gives a general summary of the outcomes associated with reciprocity in the literature, one needs to review the different approaches researchers have used to conceptualize and measure reciprocity. Several types of reciprocities are identifiable, and noteworthy. One of the earliest examples of consideration of multiple reciprocities was described by Sahlins (1965). He asserted that reciprocity can be negative (i.e. to obtain something for nothing), balanced (i.e. equality within the exchange, and payment provided swiftly), or generalized (i.e. an expectation of material return, but not governed by specific time limit, quantity, or quality of the return). By and large though, reciprocity between family members is also generalized across the life span and need not rely upon equivalent or specific exchanges of services or goods (Horwitz, Reinhard, & Howell-White, 1996). Finch and Mason (1993) noted that in kin relationships there can be differential ability to reciprocate specific services and support, but balance is maintained by exchanging different types support.

The idea of alternative reciprocities in family relationships is especially relevant to caregiving scenarios, as a care recipient may be unable to provide material support, but can maintain a balanced relationship by offering other alternatives such as providing companionship and symbolic expressions of support (Horwitz, Reinhard, & Howell-White, 1996). This view was reported in a study of caregiving in families with seriously mentally ill members. Horwitz et al., found that the reciprocity of the care recipients was mostly symbolic (i.e. expressing affection, providing companionship, participation in family activities), rather than material (i.e. financial assistance), or instrumental (i.e. providing help with chores). Horwitz and colleagues assert that the fundamental aspect of reciprocity is that the caregiver and care recipient actually perceives that the other is doing all they can to maintain a balanced relationship, while taking into account the constraints of each others' abilities. An intriguing finding was that there was a positive association between the amount of symbolic reciprocity given by the care recipient, and the amount of caregiving aid that the caregiver provided. Thus, when the opposite occurs, when low reciprocity is associated with less caregiving aid, this may have a valuable practical utility as a potential warning indicator that the caregiving relationship is under severe stress, or may soon collapse. Such a measure may help to prioritize services to distressed caregivers.

Measurement issues.

In reviewing the literature, it is apparent that there have been studies that have used the term reciprocity, as a global measure, and then there have been others where reciprocity has been subdivided as a type, i.e. symbolic, emotional, tangible, etc. There have been studies that have utilized as little as one, but as many as 4 statements, that are quantifiably assessed using Likert-type ratings. Others have used actual comparisons to what has been given, compared to what has been received – this tends to work well if there are more rigidly defined roles (e.g. in occupational settings), but does not work as well in caregiving scenarios. One must acknowledge that the very caregiving function itself results in differential expression of reciprocity, that is already acknowledged by caregivers (who are aware of the limitations that their care recipients have) and (depending on their level of awareness and cognitive ability) the care recipients themselves.

The measurement of reciprocity has not yet seen the development of a widely validated and extensively used scale. This is because reciprocity has been measured in a variety of contexts that have not been generally conducive to translational adaptation. For instance occupational reciprocities are very different from more intimate social relationships, including caregiving. In some cases, as reciprocity is seen as an assessment of balance of exchanges, it has often been measured by a tally of these exchanges.

One scale that has been developed is the Caregiver Reciprocity Scale (Carruth, 1996). The Caregiver Reciprocity Scale (CRS) is described as being able to measure the collective expression of exchanges and balance in the transactions between a caregiver and their elderly parent, over 4 subscales. There are a total of 22 items, scored on a 5 point Likert scale. The subscales are warmth and regard, love and affection, intrinsic rewards of giving, and balance of other family relationships within family caregiving. Warmth and regard measures exchange of emotions including empathy, esteem, gratitude, care, and concern. Love and affection measures the caregiver's perception of love in the relationship. The intrinsic rewards of caregiving scale assesses the motivation of the caregiver to perform their duties, including their willingness to sacrifice to benefit their care-recipient. The balance scale measures whether the caregiver feels they are balancing the requirements of their caregiving role, with their other commitments in other family relationships. The CRS is the only validated scale that has been used in other studies of caregiver reciprocity (e.g. Reid, Moss, & Hyman, 2005). It has also undergone revision, i.e. Caregiver Reciprocity Scale II (Carruth, Holland, & Larsen, 2000).

However, due to concern regarding two of the subscales being conceptually similar, this measure will not be used in the proposed study. The subscales "warmth and regard", and "love and affection", can both be considered to reflect positive emotional states. Correlation between "warmth and regard" and "love and affection" was reported to be 0.79, and the authors concede that only partial discriminant validity was demonstrated between these two factors when testing the measure. Furthermore, expectation of love can differ if the care recipient is a spouse, sibling or a child of the caregiver. Undoubtedly there is an emotional content to reciprocity, but this scale over-emphasizes it. Balance within other family relationships is not actually a direct measurement of reciprocity, reciprocity is strictly exchanges in two parties, not what is given and received to others in another relationship. In those relationships external to the caregiver/care recipient relationship, reciprocal adjustments are probably made, because of knowing the demands of the primary caregiving role. Furthermore, there is no assessment of more tangible forms of reciprocity. As this scale was designed for caregivers of elderly parents, it might be assumed that they have functional limitations that prevent tangible reciprocity. However, for caregivers of adult children with ID, tangible reciprocities have been reported to exist (Heller, Miller, & Factor, 1997). Therefore the CRS and CRS II scales were not utilized as a measure of reciprocity in this study.

Reciprocity and Intellectual Disability Caregiving

As referred to earlier, reciprocity has not been well investigated in caregivers for people with ID. To recap, the Heller, Miller, and Factor study, (1997), did find that "support from adult child with mental retardation" accounted for significant variance in caregiving satisfaction. Heller et al. stated that adults with ID provided their parents with an enduring source of companionship and constructive help with household tasks.

Given this finding, it is surprising that reciprocity has not received further research attention. Reciprocities can be reported on objective recordable behaviors, - i.e. tangible reciprocities (i.e. help with household chores). The more "emotional" reciprocities (i.e. affection displayed) are more subjective, a personal perception rather than the observable "tangible" type. It would be intriguing to investigate the level of congruence between responses given by dyads of caregivers and care recipients, particularly with the tangible reciprocities, though this was beyond the scope of this present study. However, perceived reciprocity reported by a caregiver still remains an interesting issue, given its apparent relationship to positive well-being. Reciprocity may be a significant coping mechanism in long term caregiving scenarios that are experienced by caregivers of adults with ID.

One question that should be addressed is whether perceived reciprocity of caregivers is associated with their sense of overall well-being, given that "support" is associated with decreasing burden and increasing caregiver satisfaction. As caregiving satisfaction and burden are quite specific domains, it was pertinent to investigate whether more global and clinically meaningful measures of physical and mental health can also be predicted by reciprocity.

Reciprocity has been investigated using various typologies as previously reviewed. Fundamentally, some of these typologies evolved from recognition of the different aspects of social support that caregivers received from others (Krause, 1995;

Krause & Borawski-Clark, 1995). Social support research has identified that emotional social support helped caregivers receive comfort and concern from others regarding their caregiving role, tangible social support is assistance from others for housework, and transport etc., and informational support is receiving problem-solving advice and information (Krause, 1995; Krause & Borawski-Clark, 1995). In reciprocal exchanges, the essential difference is that such support is both given and received, whereas social support is concerned with assistance provided to the caregiver, with no emphasis on what the caregiver does in return (if anything) for the person providing the support.

Although reciprocity has been categorized in several ways, there are two types (drawn directly from the literature on social support) that are particularly relevant to ID caregiving scenarios, when the respective abilities of the caregiver and care recipient can vary widely due to communication difficulties, sensory impairment, physical disabilities, and psychological/behavioral issues. These are emotional reciprocity (e.g. providing companionship, displays of affection), and tangible reciprocity (e.g. help with household tasks). It is also appealing that these typologies fit well within the stress process models as different types of coping resources. Emotional reciprocity, which is a subjective perception, may be considered an internal coping resource, whereas tangible reciprocity, can be considered an external coping resource. As previously mentioned, internal resources are inherent within the person themselves, whereas external resources are accessed by changing the environment or utilizing assistance from others. Tangible reciprocity actually provides a mechanism for the caregiver to receive assistance. The fact that it is from the care recipient does not negate its utility or diminish its impact as an "external resource".

There is not only the question of whether reciprocity might be a possible predictor for global health outcomes, but are different types of reciprocity (i.e. emotional versus tangible) differentially associated with caregiver well-being? Is it that in cases where the care recipient is highly dependent for example, and tangible tasks are unlikely/unable to be reciprocated, does an increase in perceived emotional reciprocation serve to counteract that imbalance? Perceived emotional reciprocity can be viewed as a form of cognitive appraisal that may ultimately be a coping mechanism of dealing with a dependent relationship over a lifespan. Alternatively, lack of reciprocity may also be viewed as a stressor. Either way, framing a caregiver's experience using reciprocity may be a future avenue for a stress alleviation intervention for improving caregiver well-being from removal of lack of reciprocity as a stressor, or as a potential form of coping mechanism. There is also another side to this research that determining the level of reciprocity will answer. It will help identify how much tangible reciprocity adults with ID provide in assisting their aging parents. Again, this is an area where little information is known. For persons with mild to moderate intellectual disability, it may be that as their parents age, the distinction between who is the caregiver/care recipient becomes decidedly blurred, and that role reversal in task allocation occurs.

Guiding Theoretical Model of Dissertation

This study was conducted with reference to the stress and coping paradigm as discussed previously. The two major aspects investigated with this study are compound caregiving and reciprocity. Emotional and tangible reciprocity, and compound caregiving are both potential aspects in coping with stress of ID caregiving that have not been widely studied.

Research Questions

This dissertation is focused on three major research questions.

Research Question 1 - Is compound caregiving associated with adverse quality of life outcomes when compared with non compound caregivers?

Hypothesis:

Compound caregiving (a chronic strain/life event) increases the likelihood of poorer quality of life outcomes (i.e. greater depressive symptomatology, reduced life satisfaction, poorer global health, and increased desire to place their care recipient in alternative residential accommodation).

Rationale: This question assesses the importance of this new variable as an independent stressor in the stress and coping paradigm.

Research Question 2 – Is increased reciprocity associated with better quality of life outcomes for ID caregivers?

Hypothesis:

Caregivers who report higher levels of reciprocity in their caregiving relationship, will be associated with better quality of life outcomes (i.e. reduced depressive symptomatology, increased life satisfaction, better global health, and decreased desire to place).

Rationale: This question investigates the utility of reciprocity as an independent predictor of caregiver quality of life.

Research Question 3 - Do these relationships between compound caregiving, reciprocity, and quality of life outcomes remain after controlling for other important variables in the stress process model?

Rationale: The incremental contribution of these variables is examined, in respect to stress and coping in ID caregivers, beyond more commonly studied measures of stressors and internal and external resources.

Chapter Three: Research Methodology

Study Design

This study utilized a cross-sectional design. Data was acquired from semi-structured interviews utilizing survey instruments. Participants who were located near to the Tampa Bay area were interviewed in their own home or another venue that was requested by the caregiver. For those located elsewhere or otherwise requested by the caregiver, the interview was conducted by telephone.

Participants

The study used a convenience sample of parent caregivers. In cases where there was more than one parental caregiver, the caregiver with primary responsibility, in terms of tasks and time spent with the care recipient, was interviewed. Parent caregivers were required to meet 3 eligibility criteria for inclusion in this study. Firstly, the minimum age of the parent caregiver was 50 years. Secondly, the minimum age of inclusion for their care recipient with intellectual disability was 18 years. Thirdly, the caregiver and care recipient needed to be co-residing. The presence of intellectual disability in the care recipient was determined by confirmation from the parental caregiver of previous medical diagnosis of ID, or receipt of Agency for Persons with Disability Services, or previous/current attendance within the special education system.

Recruitment Procedure

The research protocol was approved by the Institutional Review Board at the

University of South Florida. Caregivers were recruited from local agencies throughout Florida, as well as networking with local parent groups, and word-of-mouth referral. Flyers describing the criteria for the study were widely distributed using a variety of sources. These included state and local internet groups, parent support groups, posting to listserves and websites, as well as physical distribution of paper flyers, and parent-to-parent referral. Caregivers were instructed in the flyer to make contact via telephone or email, to register their interest, at which point their eligibility for the study was confirmed. Further details regarding the study were then discussed, and for those who were interviewed in their home, an appointment was made. The in-home interviews began with discussion of the informed consent procedure, and subsequent signing of the informed consent form, followed by the interview itself. In most cases, the interviews averaged approximately 1.5 hours. For those who were interviewed by telephone, the informed consent form was explained, and then a package was mailed which included informed consent forms and also a copy of the survey interview. Upon return of the signed informed consent, the caregivers were then contacted for an appointment at a convenient time. As the interview was conducted over the telephone, and the caregivers already had the questions, usual time of the telephone interviews was 45 minutes.

Measures

Independent Variables

The following variables were included to provide detailed descriptive and demographic information of the caregivers to enable thorough analysis of the overall sample characteristics. Care recipient demographic and descriptive information that have been previously associated with increasing stress and burden in ID caregivers were incorporated, so that they could be controlled for in subsequent analyses.

Caregiver demographic information.

Each participant was asked to provide details regarding their age (in years), gender, income (combined household income), education (years in education), ethnicity, and current comorbidities. Comorbidity was assessed using the self-report comorbidity scale developed by Bayliss, Ellis, and Steiner (2005). This scale has a checklist of 23 chronic medical conditions and determines both the presence of the condition, as well as a 5 point Likert-type scale to report the level of interference each of these conditions had on regular daily activities. The participants were specifically asked whether a medical doctor has ever diagnosed them with each health condition. Examples include congestive heart disease, diabetes, osteoarthritis, osteoporosis, cancer, and stroke. A total count of comorbid conditions, as well as comorbidity interference scores were obtained.

Care recipient demographic information.

Age, gender, intellectual disability level (i.e. mild, moderate, severe, profound, as previously determined by government agency/ psychological educational testing), and ID etiology were noted. Challenging behaviors were assessed using the 8 item Maladaptive Behavior subscale, of the Scales of Independent Behavior-Revised (Bruininks, Woodcock, Weatherman, & Hill, 1996). For this study, the total number of challenging behaviors present were noted, as was the total sum score of the severity of each behavior as a problem specifically perceived by the caregiver. Functional ability of the care recipient was measured using the widely used Activities of Daily Living (Katz, Ford, & Moskowitz, 1963), and Instrumental Activities of Daily Living (Lawton & Brody, 1969).

These aspects are primary caregiving stressors that have previously been identified with increasing caregiver stress. Activities of Daily Living (Katz, Ford, Moskowitz, 1963) were assessed by 6 items with yes/no responses to questions that determine whether functions such as bathing, dressing, and feeding oneself, are able to be done independently. "Yes" responses score 1 point, "no" responses score zero. The total score can range from 0-6, with lower scores indicating that greater assistance is required with activities of daily living. The Instrumental Activities of Daily Living scale (Lawton & Brody, 1969) is comprised of 8 items that assessed the level of independence in a variety of areas including use of telephone, shopping, using transport, ability to manage finances etc. The total scores can range from 0-8, with a lower score indicating less independence.

Coresidency status.

The caregiver was asked if their care recipient had continuously lived with them since birth. There were only two occurrences where the care recipient was reported to have had lived apart from the caregiver, one was prior to adoption (at a young age), the other occasion was for less than 2 years in a care recipient who was aged 30+. Therefore for the purposes of this study, it was established that the caregivers had all been lifetime caregivers.

Independent Variables to Investigate Research Hypotheses

Compound caregiver status.

Current compound caregiving status was determined by caregiver response to the question "Do you currently have caregiving tasks and responsibilities to another family member, other than your child?" This definition of compound caregiving was purposefully broad, as to capture the full range of possible multiple caregiving roles that may be undertaken. Compound caregivers includes sandwich caregivers (i.e. primary caregiver who is now responsible for older adults), but also included primary caregivers looking after any other family member, including another of their own children or grandchildren with special needs. In addition, the family relationship of who the compound caregiving duties were being provided for, and the duration of time that these simultaneous caregiving roles had existed, was noted. The caregiver was also asked approximately how many hours per week they spent undertaking the additional compound caregiving duties, and the major health condition or needs that is the reason that caregiving duties are undertaken. Previous history of compound caregiving periods were also noted, including which family member was cared for, and the major health issue or disease that resulted in the caregiver undertaking. All caregivers (i.e. both compound and non-compound caregivers) were asked if they anticipated any future caregiving responsibilities for other family members in the future.

Reciprocity.

This dissertation study used its own measure of reciprocity. This was assessed by a questionnaire that was comprised of 12 items. The scale is based on the premise that reciprocity is an exchange, so the scale must allow for the caregiver to assess each item in terms of giving to and receiving from the care recipient. Questions regarding what the caregiver felt they gave to the relationship were asked earlier in the interview, and questions asking what they felt they received were administered at the end so as to minimize the potential of socially desirable responses. The items were also chosen to be parallel across both caregiver and care recipient.

The emotional reciprocity items were:

1a) How much companionship (i.e. spending quality time together) do you give your care recipient?

1b) How much companionship (i.e. spending quality time together) do you receive from your care-recipient?

2a) How much verbal (e.g. saying "I love you") and non-verbal expressions (e.g. smiles, happy vocalizations) of positive emotion do you give your care recipient?

2b) How much verbal (e.g. saying "I love you") and non-verbal expressions (e.g. smiles, happy vocalizations) of positive emotion do you receive from your care recipient?

3a) How much physical expression of affection (e.g. hugs, kisses) do you give your care recipient?

3b) How much physical expression of affection (e.g. hugs, kisses) do you receive from your care recipient?

The tangible reciprocity items were:

4a) How much help do you give your care recipient with laundry-related tasks (e.g. placing dirty clothes in laundry basket, using washing machine/dryer, putting away clean clothes in drawers/closet, changing bed linen)?

4b) How much help do you receive from your care recipient with laundry-related tasks (e.g. placing dirty clothes in laundry basket, using washing machine/dryer, putting away clean clothes in drawers/closet, changing bed linen)?

5a) How much help do you give your care recipient with cleaning tasks around the home (e.g. vacuuming, dusting and polishing, general tidying up, help with yard work)?

5b) How much help do you receive from your care recipient with cleaning tasks around the home (e.g. vacuuming, dusting and polishing, general tidying up, help with yard work)?

6a) How much help do you give your care recipient with preparing drinks, snacks, and meals (e.g. preparing sandwiches/cereals, cooking, using microwave, toaster, kettle, dishwasher, stovetop/oven, making hot/cold drinks)?

6b) How much help do you receive from your care recipient with preparing drinks, snacks, and meals (e.g. preparing sandwiches/cereals, cooking, using microwave, toaster, kettle, dishwasher, stovetop/oven, making hot/cold drinks)?

The questions were answered with a 5 point Likert-type scale, with responses to either how much they gave/received for each item, namely - none (1), some (2), quite a bit (3), a lot (4), a great deal (5). Initially four scores were derived – emotional given, emotional received, tangible given and tangible received. Overall difference scores for both emotional and tangible reciprocity were calculated by subtracting the total received from the total given. The range of each scale was -12 to +12, with 0 begin balanced reciprocity. Negative scores indicate that more is received than given indicating a relative advantage in reciprocity, and positive scores indicate more is given than received, i.e., relative disadvantage. Cronbach's alpha for emotional reciprocity was .75, and .93 for tangible reciprocity.

Outcome Measures

There were five variables used as outcome measures; global physical health, global mental health, depressive symptomatology, life satisfaction, and desire to place. Global physical and mental health was important to ascertain whether stress and burden of caregiving actually impacts physical and mental functioning. Subjective perceptions of health are particularly useful especially in older adults, where it has been found to be a strong prospective predictor of mortality, above and beyond the effects of other variables that assess health (Benyamini & Idler, 1999). Depressive symptomatology and life satisfaction ensured that the study investigated positive and negative aspects of psychological well-being. This is important as positive and negative psychological states are not considered to be mere inverse functions of each other. They do have different antecedents, functions, and correlates, that require that they are independently measured (Lawton, 1983; Diener & Emmons, 1984; Watson & Tellegen, 1985). Finally desire to place was important to determine from both a clinical and strategic planning and policy standpoint, as it indicates the level to which caregivers are actually considering ceasing their caregiving duties.

Global physical and mental health.

Overall physical and mental health of the caregiver was measured using the Medical Outcomes Study-Short Form (SF-36). This is a 36-item instrument and has been widely validated as a summary measure of health-related Quality of Life (Ware & Sherbourne, 1992). The SF-36 has two major subscales, the physical component summary (PCS) and mental component summary (MCS), which are the scales used to assess global physical health, and global mental health in this study. Each component scale is derived from 4 minor subscales. For the PCS, the subscales are physical functioning, role-physical, bodily pain, and general health perceptions. For the MCS, vitality, social functioning, role-emotional and mental health are individually assessed. Each subscale is scored from 0 to 100 with 100 being the most favorable score (e.g., higher scores indicate better functioning for physical functioning, social functioning, role-emotional, and general health and less pain and limitations for the limitation subscales). Scores are transformed and are reported in a standardized t-score metric, i.e. mean = 50, standard deviation ± 10. For both PCS and MCS, a mean of 50 is interpreted as average health status for the overall US population. Cronbach's alpha for the PCS was .91, and .87 for the MCS.

Depressive symptomatology.

Depressive symptomatology was assessed by the 20-item version of the Center for Epidemiologic Studies-Depression Scale (Radloff, 1977). Participants were asked to report how often various feelings or behaviors were experienced during the past week, and their responses were rated on a 4-point likert scale. The categories are presented in the following order: "rarely or none of the time", "some or a little of the time", "much of the time" and "most or all of the time". Scores can range from 0 (indicating no depressive symptoms) to 60 (indicating severe depressive symptoms). Scores of 16 or higher on the Center for Epidemiologic Studies-Depression Scale (CES-D) are typically viewed as clinically significant and evidence of probable depression (Andresen, Malmgren, Carter, & Patrick, 1994). Cronbach's alpha was .89 in the present study.

Life satisfaction.

The Life Satisfaction Index – Z is the 13 item short form version of a measure designed to indicate levels of self-perceived morale and general life satisfaction (Wood, Wylie, & Schaefor, 1969). Participants were asked whether they agree, disagree or unsure one way or the other, to statements including "I am just as happy as when I was younger" and "most of the things I do are boring or monotonous". Scores range from 0 – 26 withhigher scores indicating greater life satisfaction. Cronbach's alpha in the present study was .77.

Desire for residential placement.

This variable was an adaptation of the Desire-to-Institutionalize Scale by Morycz (1985). Caregivers were asked "In respect to seeking alternative residential placement for your son/daughter, which of the following best describes you? A continuum of responses ask the caregiver the level to which they have either considered, discussed with their son/daughter the possibility, or with other family members, through to actively seeking it, and making steps to finding alternative residential placement. Scores range from 1 – 6, with 6 indicating greatest desire to place. In order to make the scale suitable and sensitive to the ID population, the wording of items was altered so as to not cause potential offense to caregivers (e.g. "institutionalization" was replaced with alternative residential accommodation, "patient" with son/daughter).

Statistical Approach

All variables were checked for normality of distribution using skewness and kurtosis indicators, and all variables were found to be in acceptable range, and required no further transformation. There were no missing data in any of the variables utilized in this study. Descriptive analyses were conducted of all independent and dependent variables. An alpha level of .05 was used for all tests of statistical significance.

The data was analyzed by initially assessing Pearson Product Moment correlations between the independent variables and the outcome measures. Any of the demographic, caregiver characteristics, or care recipient characteristics that were associated with the outcome measures were included and controlled for in the subsequent hierarchical regression analyses. The specific statistical method to answer each research question is detailed below.

Research Question 1 – Is compound caregiving associated with adverse quality of life outcomes when compared with non-compound caregivers?

Independent sample t-tests were conducted to compare sample characteristics of current compound caregivers with non-compound caregivers. For categorical variables, Pearson's chi-square tests were used. To specifically assess quality of life differences between the two groups, independent sample t-tests were conducted in global health functioning (MCS/PCS), depressive symptomatology, life satisfaction, and desire for residential placement.

Research Question 2 – Is increased reciprocity associated with better quality of life outcomes for ID caregivers?

Pearson Product Moment bivariate correlations were conducted between tangible reciprocity and emotional reciprocity and the quality of life outcomes, global health functioning (MCS/PCS), depressive symptomatology, life satisfaction, and desire for residential placement.

Research Question 3 - Do these relationships between compound caregiving, reciprocity and quality of life outcomes remain after controlling for other important variables in the stress process model?

The predictive value of compound caregiving status and reciprocity was assessed using hierarchical multiple regressions. The stress-coping paradigm provided the guidance for the regression models. Firstly, demographic and significant caregiving stressors at univariate level were entered. Secondly, compound caregiving, an example of a chronic life event/strain was added to the model. Reciprocity was then added last to each model.

Chapter Four: Findings

Sample Characteristics

Caregiver Characteristics

A total of 91 caregivers participated in the study. The caregivers averaged just over 60 years of age but there was considerable range. There were 47 caregivers who were interviewed over the telephone, and 44 who were interviewed in their home. The majority of this sample (78 caregivers) were recruited from throughout Florida, in particular the greater Tampa Bay area, Miami-Dade, Pensacola, and Fort Myers area. However, due to internet recruitment, caregivers from other states were also recruited. This included 5 from New Jersey, 4 from Georgia, and 1 each from Maryland, Oklahoma, South Dakota, and Nevada. The caregivers were predominately white and female. Those who identified in the non-white category included 4 caregivers who were African American/Black, 2 who were Hispanic/Latino, and 1 Pacific Islander. The caregivers were highly educated, with the average participant having completed a few years of college education.

With respect to compound caregiving status, over a third of the sample were currently undertaking compound caregiving duties. Furthermore there were many who if not current a compound caregiver, had been in the past. Regarding the possibility of becoming a future compound caregiver for other family members (regardless of their present compound caregiving status), approximately one third responded that they were anticipating that would be a likely scenario. Table 4.1 presents both caregiver and care recipient characteristics.

Table 4.1. Sample Caregiver and Care Recipient Characteristics (N = 91).

	Mean or %	SD	Range
Caregiver Characteristics			
Demographic			
Age (years)	60.8	8.5	50 – 92
Education (years)	15.1	2.4	12 – 22
Gender (Female)	91%		
Race/Ethnicity (White)	92%		
Household Income (>$50,000)	78%		
Health			
Total Comorbidities	5.1	2.9	0 – 13
Comorbidity Interference	9.2	7.9	0 – 35
Caregiving			
Total caregiving hours per week	39.4	21.3	7 – 88
Compound Caregiver Now (Yes)	37%		
Compound Caregiver Ever (Yes)	68%		
Anticipated Future Caregiving (Yes)	34%		
Care Recipient Characteristics			
Demographic			
Age	29.7	9.5	18 – 54
Gender (Female)	40%		
Health and Functional Abilities			
Activities of daily living (ADL)	4.6	2.1	0 – 6
Instrumental ADL's	2.3	1.7	0 – 6
Intellectual disability level	2.0	0.8	1 – 4
Challenging behaviors (total)	2.9	2.5	0 – 8
Challenging behaviors (severity)[a]	6.1	6.0	0 – 26

[a] N = 63

Care Recipient Characteristics

There were 55 male and 36 female care recipients, and their mean age was 29 years, with considerable variation. Etiology of intellectual disability was Down syndrome (28), Unknown etiology (18), Autism Spectrum Disorders (15), Others (14), Cerebral Palsy (12), Fragile X (4). Within the category of "others" there were two with Tuberous Sclerosis, one care recipient with Prader-Willi Syndrome, another with Williams syndrome. Unknown etiology refers to intellectual disability of unknown origin after a formal diagnosis of a chromosomal disorder, or physical condition/disease precipitating the intellectual disability has not been medically determined. Overall, this was quite a diverse sample in terms of etiology. The majority of care recipients had a moderate level of intellectual disability. Overall, 27 care recipients were in the mild ID category, 44 in the moderate category, 12 were severely intellectually disabled, and 8 had profound ID. The means, standard deviations, and ranges for all care recipient demographic details are also detailed in Table 4.1.

The quality of life outcome measures for the entire sample are presented in Table 4.2. For PCS and MCS, a mean of 50 indicates average health status in the general US population. Overall, caregivers reported poorer physical health, as the mean for the PCS scores were over 0.5 of a standard deviation lower. Global mental health was also reported to be slighter lower (.2 of a standard deviation below).

Table 4.2. Mean and Standard Deviation for Main Outcome Measures (N=91)

Outcome Measure	Mean	Standard Deviation	Range	Range of Instrument
Life Satisfaction (Life Satisfaction Index)	17.38	6.57	3 - 26	0 - 26
Depressive Symptomatology (CES-D)	11.98	9.45	0 - 40	0 - 60
Global Physical Health (PCS of SF-36)	43.61	10.50	17.1 - 65.9	0 - 100
Global Mental Health (MCS of SF-36)	48.29	11.36	17.1 - 64.5	0 - 100
Desire to place	2.9	1.75	1 – 6	1 - 6

Compound Caregiver Characteristics

Table 4.3 presents the major types of relationships between compound caregivers and their care recipients. The largest categories of compound caregiving recipients were for mothers, fathers, and spouses. Table 4.3 also presents the major health issue of the compound caregiving care recipient, the most common being Alzheimer's disease.

In terms of duration of caregiving, there were several compound caregivers in sample that had been compound caregivers for many years. These outliers inflated the mean duration of caregiving to 77 months, with a range from 3 – 504 months. However, the median duration was 36 months, and the modal response was 24 months.

Table 4.3. Compound Caregivers' Care Recipient Relationship and Health Issue (N = 34)

Relationship	*N*	%
Mother	13	(38.2%)
Father	4	(11.8%)
Spouse	4	(11.8%)
Sibling	3	(8.8%)
Aunt/Uncle	3	(8.8%)
2nd Child with Intellectual Disability	3	(8.8%)
Mother in Law	2	(5.9%)
Grandchild with Medical Needs	1	(2.9%)
Major Health Issue		
Alzheimer's disease	7	(20.6%)
Elderly Frail	4	(11.8%)
Advanced Macular Degeneration	4	(11.8%)
Cardiovascular Disease	4	(11.8%)
Intellectual Disability	4	(11.8%)
Parkinson's disease	2	(5.9%)
Cancer	2	(5.9%)
Chronic Mental Disorder	2	(5.9%)
Hip Fracture/Replacement	2	(5.9%)
Stroke	1	(2.9%)
Diabetes	1	(2.9%)
Post-Operative Convalescence	1	(2.9%)

Comparisons between Compound and Non-Compound Caregivers

Using independent sample t tests (Table 4.4), and Pearson's Chi square analyses (Table 4.5) compound caregivers' demographic variables, caregiving variables, and care recipient characteristics, were compared with non-compound caregivers. Overall, compound caregivers were very similar to their non-compound caregiving counterparts, with three exceptions; caregiving hours, caregiver ethnicity, and care recipient gender. With respect to the time spent in their primary caregiving role, there was no appreciable difference, but when the additional compound caregiving hours were added, compound caregivers weekly tally increased to around 52 hours. The compound caregivers were all

Table 4.4. T-Test Analyses - Compound versus Non-Compound Caregivers.

	Compound Caregiver		Non-Compound Caregiver			
	M	SD	M	SD	df	t
Caregiver Characteristics						
Demographic						
Age (years)	58.8	7.9	61.1	8.9	89	1.22
Education	14.7	2.29	15.4	2.53	89	1.26
Health and Caregiving						
Total Comorbidities	4.79	2.96	5.42	3.01	89	.97
Comorbidity Interference	8.68	8.66	9.52	7.47	89	.494
Total Caregiving hours	38.66	20.82	39.84	21.89	89	.253
Caregiving hours + CCG hours[a]	51.60	26.34	39.84	21.88	89	-2.30*
Care Recipient Characteristics						
Demographic						
Age	28.38	8.91	30.51	9.80	89	1.03
Health and Functional Abilities						
Activities of daily living (ADLs)	4.88	1.93	4.47	2.12	89	-.92
Instrumental ADLs (IADLs)	2.64	1.91	2.21	1.52	89	-1.20
Intellectual disability level	1.94	.91	2.05	1.52	89	.577
Challenging behaviors (total)	3.44	2.57	2.68	2.42	89	-1.04
Challenging behavior (severity)	5.62	6.73	6.47	5.62	61	.540

[a] CCG = Compound caregiving * $p < .05$ (2-tailed).

white. Regarding gender of the care recipients, the distribution of compound care recipients was fairly equal, but for non-compound caregivers male care recipients predominated. A final chi square analysis was conducted on the etiology of intellectual disability between the groups and no significant difference arose.

Table 4.5. Chi Square Analyses - Compound versus Non-Compound Caregivers.

	Compound Caregiver	Non-Compound Caregiver			
	%	%	N	df	χ^2
Caregiver Characteristics					
Demographic					
Gender (Female)	91%	91%	91	1	.00
Ethnicity (White)	100%	86%	91	1	5.23*
Income (>$50,000)	82%	75%	91	1	.59
Care Recipient Characteristics					
Demographic					
Gender (Male)	47%	68%	91	1	4.06*

* $p < .05$ (2-sided).

Research Question 1- Results

Was compound caregiving associated with adverse quality of life outcomes when compared with non-compound caregivers?

The results of the independent samples t-tests between compound caregivers and non-compound caregivers are presented in Table 4.6. Compound caregivers did not statistically differ from non-compound caregivers in four of the outcome measures; life satisfaction, depressive symptomatology, physical and mental health. However, their desire to place their primary care recipient with intellectual disability was significantly higher than their non-compound caregiving counterparts. The top 2 items of the scale, i.e. "I am likely to seek residential placement", or "I have made actual steps towards placement", revealed that 32% of compound caregivers had responded that they were currently at that level, compared with only 10% of the non-compound caregivers.

Table 4.6. Compound versus Non-Compound Caregivers in Outcome Measures.

	Compound Caregiver		Non-Compound Caregiver			
	M	SD	M	SD	df	t
Life Satisfaction (Life Satisfaction Index)	17.05	5.81	17.58	7.03	89	.36
Depressive Symptomatology (CES-D)	10.94	9.30	12.61	9.65	89	.85
Physical Health (PCS of SF-36)	44.51	11.28	43.09	10.07	89	-.62
Mental Health (MCS of SF-36)	47.66	11.22	48.67	11.53	89	.41
Desire to place	3.61	1.72	2.49	1.63	89	-3.11**

** p< .01 (2-tailed).

Research Question 2 - Results

Was increased reciprocity associated with better quality of life outcomes for ID caregivers?

The mean score for tangible reciprocity was 4.7, and the range was -9 to 12. The mean for emotional reciprocity was 1.6, and the scores ranged from -8 to 12. Therefore in this sample, caregivers reported greater disadvantage of reciprocity, i.e. they felt that they gave more to the relationship in both emotional and tangible terms, than was received. However, the disparity in tangible reciprocity was greater than emotional reciprocity.

The bivariate correlations between all five dependent variables and emotional and tangible reciprocity are presented within the correlation matrix in Table 4.7. Bivariate correlations between tangible reciprocity and the outcome variables resulted in several significant associations. Higher relative disadvantage in tangible reciprocity was associated with increased depressive symptomatology scores. Higher relative disadvantage in tangible reciprocity was also associated with poorer global mental heath (MCS). Higher relative disadvantage in tangible reciprocity was also associated with reduced desire to place scores. Emotional reciprocity was not significantly correlated with any of the outcome measures. Life satisfaction and physical health were not significantly correlated to either tangible or emotional reciprocity scores.

Research Question 3 - Results

Did the relationships between compound caregiving, reciprocity, and quality of life outcomes remain significant after controlling for other important variables in the stress process model?

For all hierarchical multiple regressions, the conventional formula for determining adequate sample size is 50 + 8M, where M represents the number of independent variables that are examined (Tabachnick & Fidell, 1996). With due consideration to the constraints of the current sample size and limiting the number of variables that were entered into the hierarchical regression analyses, only predictor variables that were significantly associated with the outcomes variables were included. Therefore, compound caregiving status was only included as a predictor variable in the regression model for desire to place. Regarding the outcome variables, as neither emotional reciprocity, tangible reciprocity, nor compound caregiver status, were found to be associated at univariate level for global physical health, and life satisfaction, further multivariate analysis was not conducted. Correlations between all study variables and outcome measures that were significant were incorporated into the regression models.

Depressive symptomatology and global mental health were both associated with total challenging behaviors and challenging behavior severity. These two variables are also highly correlated, ($r = .79$, $p < .01$). To reduce the possibility of multicollinearity, and to help restrict the number of predictor variables due to the sample size, only the variable challenging behavior severity was entered hierarchical regression analyses. This was because the severity score had slightly higher correlations with the outcome measures than the total number of challenging behaviors present.

Similarly, with comorbidity interference and total comorbidities, the correlation between these variables was high ($r = .76$, $p < .01$), and therefore as both variables were associated with depressive symptomatology, comorbidity interference was entered into the regression, as the correlation was more significant.

Table 4.7 Correlation Matrix for All Study Variables

	1	2	3	4	5	6	7	8	9	10	11	12	13	14	15	16	17	18	19
1 Desire For Placement		14	01	- 09	01	05	04	- 02	- 09	- 14	- 20	02	.41**	.34**	- 19	06	- 18	03	-.23**
2 Global Physical Health			- 11	- 16	.26*	-.38**	12	14	-.65**	-.66**	- 10	-.36**	02	- 07	- 95	11	06	03	- 04
3 Global Mental Health				70**	.48**	.28**	03	03	- 16	-.31**	-.29**	.31**	10	.23*	02	-.52**	-.57**	- 05	-.24*
4 Depressive Symptomatology					-.59**	- 11	01	- 18	.22*	.41**	.29*	- 14	- 11	-.21*	- 01	.32**	.42**	14	.29**
5 Life Satisfaction						07	.28**	18	-.24*	-.38**	-.24*	03	08	12	- 03	-.26*	-.35**	- 14	- 18
6 Caregiver Age							- 13	-.45**	.33**	13	- 11	.78**	12	04	- 03	-.39**	-.27*	- 00	- 07
7 Caregiver Education								.25**	- 16	- 05	-.26*	-.33**	08	12	- 18	06	- 02	04	- 14
8 Caregiver Income									- 18	- 09	- 08	-.32**	- 11	10	- 04	16	11	13	- 01
9 Total Comorbidities										.76**	10	.35**	02	05	- 01	- 04	06	- 08	03
10 Comorbidity Interference											20	07	02	- 00	10	17	.29*	05	10
11 Total Caregiving Hours												- 15	-.54**	-.50**	.47**	.34**	21	19	.45**
12 Care Recipient Age													10	07	03	-.51**	-.50**	- 08	- 05
13 ADLs														.63**	-.63**	- 19	- 18	- 16	-.64**
14 IADLs															-.61**	-.36**	-.34**	- 06	-.69**
15 Intellectual Disability Level																08	06	.26*	.57**
16 Challenging Behaviors Total																	.79**	10	.31**
17 Challenging Behavior Severity																		03	23
18 Emotional Reciprocity																			20
19 Tangible Reciprocity																			

*p<.05, **p<.01, ***p<.001

Regression Model – Depressive Symptomatology

For depressive symptomatology, (see Table 4.8), the regression model accounted for 37.5% total variance in the scores. The demographic and caregiving stressors predicted a significant amount of variance, with higher levels of comorbidity interference being a significant individual predictor variable in the model. The addition of tangible reciprocity as a coping resource, did not predict any additional significant variance in the model, and comorbidity interference remained the only significant predictor in the combined model.

Table 4.8. Hierarchical Regression Analysis Predicting Depressive Symptomatology

Variables	Standardized Beta Coefficients	R^2	ΔR^2
Model 1: Demographics & Caregiving Stressors			
Instrumental ADLs of CR	-.155		
Challenging Behavior Severity of CR	.231		
Comorbidity Interference	.401***		
Total Caregiving Hours	.058	.350	.350***
Model 2: Model 1 + Reciprocity (Coping Resources)			
Tangible Reciprocity	.206	.375	.025

***$p<0.001$ (2-tailed).

Regression Model – Global Mental Health

In the hierarchical regression model to predict global mental health (see Table 4.9), 40.4% of the variance in the scores was accounted for. Significant variance was again predicted by demographic and caregiving stressors. In particular higher scores in challenging behavior severity, and higher comorbidity interference were significant predictors of global mental health. However, the addition of tangible reciprocity, as a coping resource, did not add any further variance to this model.

Table 4.9. Hierarchical Regression Analysis Predicting Global Mental Health

Variables	Standardized Beta Coefficients	R^2	ΔR^2
Model 1: Demographics & Caregiving Stressors			
Caregiver Age	.109		
Care Recipient Age	.022		
Instrumental ADLs of CR	.001		
Challenging Behavior Severity of CR	-.437**		
Comorbidity Interference	-.267*		
Total Caregiving Hours	-.072	.404	.404***
Model 2: Model 1 + Reciprocity (Coping Resource)			
Tangible Reciprocity	-.004	.404	.000

*p<.05 (2-tailed), **p<.01(2-tailed), ***p<0.001 (2-tailed).

Regression Model – Desire to place

The regression model to predict desire for residential placement accounted for 25.9% of the variance in scores (Table 4.10). Caregiving stressor variables accounted for significant variance, with higher levels of activities of daily living being a significant individual predictor. When compound caregiving was added to the model as a chronic stressor, significant additional variance was also added, and compound caregiving status was also a significant predictor variable. Therefore, those who were compound caregivers were more likely to desire residential placement for their primary care recipient. When tangible reciprocity (a coping resource) was added, there was no significant increase in the variance explained. Activities of daily living, and compound caregiving status remained significant predictors in the model.

Table 4.10. Hierarchical Regression Analysis Predicting Desire For Placement

Variables	Standardized Beta Coefficients	R^2	ΔR^2
Model 1: Demographics & Caregiving Stressors			
Activities of Daily Living (CR)	.326*		
Instrumental ADLs (CR)	.131	.177	.177***
Model 2: Model 1 + Compound Caregiving (Chronic Stressor)			
Current Compound Caregiver	.270**	.249	.071**
Model 3: Model 2 + Reciprocity (Coping Resource)			
Tangible Reciprocity	.149	.259	.010

*p<.05(2-tailed), **p<.01(2-tailed) ***p<0.001 (2-tailed).

Chapter Five: Discussion

As this study was exploratory, it yielded some interesting findings that were not central to the three specific research questions addressed in this dissertation study. Therefore, the discussion will first focus on exploring the findings with regard to the study's hypotheses. The second section will discuss other relevant findings. Limitations of the study and future directions will then be explored.

Study Findings

The first research question was to determine whether compound caregivers had poorer quality of life outcomes compared with their non-compound caregiving counterparts. In the present study, compound caregiving was not found to be associated with poorer outcomes in terms of life satisfaction, depressive symptomatology, physical or mental health as hypothesized. However, the hypothesis that compound caregivers were more likely to have a greater desire to seek alternative residential placement for their care recipient, was supported. This may well be that the very act of being a compound caregiver has bestowed upon these parents the urgency of addressing "what will happen to my son/daughter if I get sick" type of scenarios. It could also be argued that they are experiencing more strain, and thus are more willing to cease their primary caregiving duties. Many more compound caregivers endorsed the items of "likely to seek" or "have made actual steps" towards alternative residential placement compared with the non-compound caregivers.

The majority of compound caregivers were looking after their parents, spouses, or in laws with chronic health conditions, so such questions may be more pressing in their own mind as well as the aging process. Therefore, it would seem reasonable to suggest that compound caregiving status does serve to galvanize the discussion about care recipients living elsewhere in some of these families.

With respect to the fact that compound caregivers' physical and mental health, depressive symptomatology, or their life satisfaction was not significantly different from the non-compound caregiver, there may be two possible explanations. Firstly, older caregivers have a lifetime of caregiving experience, and become experts at their caregiving role (Haley & Perkins, 2004). Therefore an additional caregiving role may not be too onerous a burden for these highly experienced caregivers. Indeed, many compound caregivers commented that it was natural that they undertook the compound caregiving roles. A lifetime of caring for a child with special needs, may certainly better equip these lifetime caregivers with the knowledge, skills, and empathy required to do so. Therefore, it may be an easier transition to a compound caregiving role, than it is for those who are not previous caregivers within the immediate family – thus a type of natural selection, indeed even self-selection process manifests.

However, in some cases it was expressed that there was an obvious expectation from others in the family that this person take on the extra caregiving role. Indeed it may be a possibility that subtle coercion may also operate - as some caregivers commented that other siblings had made remarks that it may be easier for the caregiver to look after Mom/Dad etc, as they were already doing it for their Son/Daughter! Sometimes, especially if the caregivers were already spending considerable time in the primary caregiving role at home, there was a perception that additional caregiving responsibilities could be more easily undertaken and be less disruptive to the caregiving routines already established.

Another possible explanation of lack of difference in present compound caregiver's quality of life outcomes compared with non-compound caregivers, is that during the compound caregiving episode, one actually becomes adjusted to the task in hand, and may minimize the physical or mental effects that may be manifesting. If one currently has caregiving responsibility for two care recipients, then it might become a coping mechanism, to suppress any thoughts of ill-heath. As one compound caregiver remarked "you just need to get on with it".

The second research question in this dissertation was to investigate whether reciprocity was associated with better quality of life outcomes for ID caregivers. Relative disadvantage in reciprocity (i.e. indicating the caregiver gives more than they receive) was found to have significant correlations with depressive symptomatology and global mental health as hypothesized. Therefore, if a caregiver feels like there is considerable negative imbalance in the exchange of duties within the caregiving relationship (e.g. that one is giving much more than one is getting), poorer mental health outcomes are associated. Relative disadvantage in reciprocity indicates that the care recipient has significant needs, thus is unable to provide tangible help to their parent in any meaningful way for maintaining the family household. This raises the question of whether the number of other able co-habitants, and tangible help available from others, needs to be ascertained in assessments of reciprocity. If a caregiver has adequate support from others, then the issue of tangible reciprocity is less important. However, if tangible help is not readily available, relative disadvantage in reciprocity in caregivers who live alone with their care recipient, is likely to assume much greater importance in its role as an external coping resource.

An interesting finding, and opposite to what was hypothesized, was the finding that relative disadvantage in tangible reciprocity was associated with a decreased desire for residential placement. It would appear in this study, that caregivers of those with greater needs are less likely to want to place their son or daughter into residential care. However, this may be that such caregivers are more committed to looking after their son/daughter with highly dependent needs rather than considering the alternatives of other residential placements. There were several caregivers in the current study with very severely or profoundly intellectually disabled children, who required considerable attention due to their associated secondary conditions (e.g. severe epilepsy, PEG feeding from dysphagia, respiratory therapy, hemiplegia and quadriplegia, congenital blindness/deafness, limited communication skills). These caregivers were quick to assert that they felt the most qualified to look after their care recipient, even when they were receiving personal companion/assistance services. Indeed, there were several caregivers that did not feel comfortable leaving their son/daughter in the company of in-home personal companions in some cases, because they felt that the care workers were not as adept in the care of their son/daughter. Others remarked that the very fact that their son or daughter was highly dependent on them, made them unsuitable candidates for alternative residential placement. The notion of the care recipient living elsewhere was asserted by some, to be a better option only for those people with intellectual disabilities, who are were more able, verbal, and independent. There was considerable apprehension from some caregivers that highly dependent care recipients do not fare as well in such settings, another possible reason that may explain the finding of relative disadvantage in tangible reciprocity reduced desire for placement in this study.

Life satisfaction or physical health was not associated with tangible reciprocity. Emotional reciprocity was not associated with any outcome measure. This was somewhat surprising given the relative importance of the primary caregiving role, that emotional benefits derived from the relationship did not manifest any significant associations. Overall, caregivers reported to feeling that emotional reciprocity was reasonably equitable.

Lack of significant results may be a result of the items the emotional reciprocity scale used. Considerable thought was given to the wording of such items, so as to take into account the wide range of communication abilities that would exist in the care recipients in this study. However, it may be the case that the items were too broad or lack the sensitivity to truly capture emotional reciprocity because of this emphasis on making the items applicable to all. Furthermore, tangible reciprocity does not incur inherent value judgments about a relationship in the same way that emotional evaluations do. Caregivers may have a found it difficult to accurately evaluate their relationship in such a manner, or else emotional reciprocity is still perceived to be very evenly balanced in circumstances where little or no reciprocity is markedly evident to the observer. That is not to deny there are those care recipients that despite physical, neurological, or behavioral issues, that may interfere somewhat in their ability to emotionally connect with others, nevertheless, rapport, and emotional exchanges can still be achieved. Oftentimes, it is just more dependent on more subtle body language, and gestures. Reciprocity, therefore seems to have some element of appraisal in it, e.g. caregivers may also incorporate a judgment that includes some consideration of the capacity of the other person to give back.

Items that may be more discriminating of emotional reciprocity might need to include more objective indicators and consider such things as time spent in mutually enjoyable activities. Fundamentally, the greatest issue may be that when emotional reciprocity is assessed between a non-intellectually disabled person and their adult son/daughter who continues to live in the parental home, there is the obvious fact that these relationships are cherished by caregivers. Thus it would be interesting to see whether emotional reciprocity between parents and adult children with intellectual disabilities are not as equal when they live apart.

Research question three investigated the utility of compound caregiving status and reciprocity as predictor variables that can account for variance in scores over all other study variables. In the present study, tangible reciprocity was not found to be a useful individual predictor. However, compound caregiving status was found to be a significant predictor of desire for residential placement, even after controlling for other predictor variables.

Increased desire for placement can have two possible reasons, there is either depletion in ones coping resources, that result in a caregivers increasing desire to discontinue the caregiving role. Alternatively, there is increased openness to the prospect of their son/daughter to living elsewhere. To err on the side of caution, it would seem reasonable with limited available resources, and state budget deficits that have necessitated long waiting lists for services to adults with intellectual disabilities, that compound caregiving roles should be an integral part of the assessment process. Compound caregiving status should definitely be a factor for prioritizing services. Even though the current study found no appreciable difference in quality of life compared with other non-compound caregivers, there remains the distinct possibility that the continued demands of compound caregiving, or multiple episodes of compound caregiving, may well lead to poorer outcomes over the long-term. Therefore, timely assistance to the primary caregiving role, in terms of additional support, may actually delay placement. This may indirectly save substantial costs that arise from providing alternative residential care, indeed, potential crisis intervention in some cases, rather than increasing in-home supports, or even respite care during compound caregiving periods.

Other Relevant Findings

It appears that compound caregiving is actually quite commonplace. As this study is one of the first to isolate compound caregivers in this manner, there was little expectation on just how many compound caregivers would be identified. Indeed, it was a possibility that if insufficient numbers within the sample had not been interviewed, that compound caregivers were to be oversampled. This proved not to be necessary. Whilst this was a convenience sample, nevertheless, 37% of the caregivers were subsequently found to be compound caregivers during their interviews. Furthermore, study recruitment materials did not specifically mention this particular aspect of the study, i.e. that there would be questions regarding caregiving for others in addition to their primary caregiving role. Furthermore, 68% were either a current or had previously been a compound caregiver. This suggests the possibility that "compound caregiving" per se, may be too common a factor for it to be predictive of many fundamental outcomes. To add further weight to this argument, irrespective of current compound caregiving duties, when asked whether any future caregiving for a family member was anticipated, 34% of the sample reported "yes". It may be that one or two episodes of compound caregiving are in fact to be expected in most circumstances. Therefore, perhaps greater attention should be focused on compound caregivers who have had multiple episodes of compound caregiving. In a case study of a compound caregiver by the dissertation author, the caregiver had 4 separate episodes of compound caregiving duties (Perkins, in press), with the fourth caregiving episode resulting in considerable distress. In the present study, questions were asked regarding the current, and up to two previous episodes of compound caregiving. The sample actually had 6 participants that were currently compound caregivers, and who had on two previous occasions been compound caregivers to other family members in the past. There were also 5 triple caregivers, i.e. had caregiving duties to their son/daughter with intellectual disabilities, and two other family members concurrently. Unfortunately, these particular caregiving groups were too small, even if combined, to allow for independent analysis to answer the specific research questions of the present study. However, to illustrate that multiple caregiving responsibilities is likely to be seriously detrimental to quality of life, with the "triple" caregivers, a quick comparison is quite illuminating. The mean depressive symptomatology score for the sample was 12.0, but for triple caregivers it was 18.2 (indicating that this was a significantly depressed group of individuals). Furthermore their life satisfaction score was only 12.4 (5 points lower than the sample mean of 17.4). Even with the obvious limitation of the small sample size - nevertheless the contrast is quite apparent. This study also highlighted the fact that, though their numbers may be small, there are caregivers who have more than one child with intellectual disabilities, and the issues of being a lifelong caregiver to two children are doubtless magnified.

With regards to emotional reciprocity, only one variable in the entire study was significantly correlated, intellectual disability level of the care recipient ($r = .26$, $p = .013$). Therefore, the greater the severity of intellectual disability, the higher the relative disadvantage in emotional reciprocity. Thus, there does seem to be some appraisal of limitations imposed through increasing communication difficulties, of what is emotionally given to what is received, but some inherent mechanism overrides the impact. Perhaps at some level there is cognitive dissonance about true emotional reciprocity that serves to protect the caregiver from feelings of inequity, and possible lack of emotional nourishment that is derived in the relationship.

Study Limitations

The present study has a number of limitations that need to be acknowledged. Firstly, this was not a randomized sample, and the convenience sampling methodology can present several biases in the results. First, it may be that healthier caregivers were more willing to participate, whereas caregivers who did not volunteer may have more adverse health outcomes. This may lead to data that suggests more positive outcomes than actually exists in the population. Another possibility is that the participants in this sample were very well educated and affluent, and thus more likely to be knowledgeable and resourceful in their own health and wellbeing, as well as accessing resources that are available to them, their primary care recipients and their compound caregiving recipients. The parent-to-parent referral that was used to recruit some participants, may also have biased the sample to include more caregivers who greatly benefit from durable support networks with other parents.

Conversely, previous research has also indicated that caregivers who are more depressed are likely to participate in studies if they are offered the alternative of in-home interview (Dura & Kiecolt-Glaser, 1990), therefore the present study may inadvertently have captured more caregivers who felt compelled to tell their story. Randomized sampling would help minimize such bias. It is important to mention that both telephone and in-home interviews were utilized in the present study, however it was necessary in order to maximize the sample size.

There are some design issues that also warrant discussion. In particular, the cross-sectional design limits interpretation to the assessment of one time point only. Longitudinal changes in caregiver well-being, especially during periods of high stress that may be experienced during compound caregiving would allow the investigation of the potentially long-lasting ramifications to health that can arise. Similarly, a diminishing sense of reciprocity over time may help to indicate periods of increasing strain and tension with the caregiving role.

This study also relied on self-reports of well-being. Though it would be advantageous to include objective clinical measurements of physical health, time and budget constraints prevented this possibility.

The sample size and comparison groups may have reduced the statistical power of the analyses conducted in this study, and it is possible that Type II errors have occurred that have obscured the detection of smaller or moderate statistically significant differences. Future studies would benefit from an increase in study participants to overcome this particular limitation.

Due to time/budget constraints, the current proposal did not utilize a non-caregiving control group. The rationale that led to this decision was that the two exploratory aspects (i.e. reciprocity and compound caregiving) would be better served by investigating a larger sample of caregivers, rather than devoting time and resources to a separate control group. The results of the present study have provided valuable pilot data that can guide more methodologically rigorous studies in the future.

Future Directions

Although limited by sample size and cross-sectional design, this study has helped to identify several avenues for future research. Compound caregiving is more common than anticipated, and the lack of finding any significant quality of life decrements compared with other caregivers might have been exacerbated by heterogeneity of the compound caregivers used in the current sample. More robust differences may be detected in compound caregivers grouped in more homogenous categories, such as those caring for parents with Alzheimer's disease, or caregivers that have had multiple episodes of compound caregiving, to those that have more than one child with intellectual disabilities. This study has highlighted the need to consider several different compound caregiving scenarios and refinement of criteria that describes compound caregiving. Other factors that may be need to be assessed are the composition of the family household (i.e. who lives with the compound caregiver), and whether the compound care recipient also lives in the family home or elsewhere. It is possible that some caregivers may live alone with two care recipients. Investigation of specific issues that can arise in compound caregiving, such as prioritizing caregiving tasks, difficulties maintaining employment, feeling stressed, feeling physically and emotionally drained, availability of adequate help, are just a few examples of pertinent problems that may arise. Further research as to the exact mechanism of why compound caregivers are more likely to seek placement is also warranted.

This dissertation found that emotional reciprocity was unrelated to quality of life. The items used for emotional reciprocity may benefit from being more objectively quantified items that focus on shared activities, and other aspects that can indicate the emotional quality of the relationship. This might include having fun and laughter together, shared hobbies, mutual friendships with others. Tangible reciprocity may benefit from inclusion of more specific chores and inclusion of how often these chores are done on a regular basis. It may be that rather than grading the reciprocity items separately as was done in this study (i.e. what the caregiver felt they gave was assessed at the start of the interview, and what they received was given at the end) – an alternative scale that one judges across a continuum of given versus received may increase the accuracy of response. Furthermore, emotional reciprocity may actually differ between parents whose son/daughter with ID currently lives in an alternative residential setting, compared with those caregivers whose son/daughter lives at home. Another interesting avenue would be to determine the level of congruence between the exchange of reciprocity that exists, as reported by dyads of caregiver and care recipient, especially in tangible reciprocities. This would also be interesting in caregiving scenarios in general, and not just limited to ID caregiving.

Overall, it is hoped that this dissertation study will have some practical utility in emphasizing the commitment that lifelong caregivers of adults with intellectual disabilities make on a daily basis to improve the quality of life of the most vulnerable in our population. It is a testament to these parents, that so many of them also assume the mantle of compound caregiver, and provide care and compassion to other members of their family when their primary caregiving role is already a substantial undertaking. Proper recognition of the extent of their caregiving roles and commitments by clinicians, service providers, policymakers, stakeholders, and indeed, in some cases, their own family members, is long overdue. It is hoped that this dissertation will, in some small way, be part of the essential groundwork that sets the stage for further large scales studies in this new area of caregiving research.

References

Adams, J. S. (1965). Inequity in social exchange. In L. Berkowitz (Ed), *Advances in Experimental Psychology* (Vol. 2, pp. 267-299). New York: Academic Press.

American Association on Intellectual and Developmental Disabilities. (2007). *Board Position Statement: Growth Attenuation Issue.* Retrieved August 31, 2008, from http://www.aaidd.org/ Policies/growth.shtml

American Psychiatric Association. (2000). *Diagnostic and statistical manual of mental disorders* (4th ed., Text Revision). Washington, DC: Author.

Andresen, E. M., , Malmgren, J. A., Carter, W. B. & Patrick, D. L. (1994). Screening for depression in well older adults: Evaluation of a short form of the CES-D. *American Journal of Preventive Medicine, 10,* 77-84.

Arias, E. (2006). *United States Life Tables, 2003* (National Vital Statistics Reports). Retrieved August 20, 2008, from http://www.cdc.gov/nchs/data/nvsr/ nvsr54/nvsr54_14.pdf

Bayliss, E. A., Ellis, J. L., & Steiner, J. F. (2005). Subjective assessments of comorbidity correlate with quality of life health outcomes: Initial validation of a comorbidity assessment instrument. *Health and Quality of Life Outcomes,* 3, 51.

Benyamini, Y., & Idler, E. L. (1999). Community studies reporting association between self- rated health and mortality: additional studies 1995-98. *Research in Aging, 21,* 392–401.

Black, M., Cohn, J., Smull, M., & Crites, B. (1985). Individual and family factors associated with risks of institutionalization of mentally retarded adults. *American Journal of Mental Deficiency, 90,* 271-276.

Blatt, B., & Kaplan, F. (1966). *Christmas in purgatory: A photographic essay on mental retardation.* Boston: Allyn & Bacon.

Braddock, D. (1999). Aging and developmental disabilities: Demographic and policy issues affecting American families. *Mental Retardation, 37,* 155-161.

Braddock, D., Hemp, R. & Rizzolo, M. C. (2005). *Trends in residential services – persons served by setting in 2004* (Coleman Institute and Department of Psychiatry, University of Colorado). Retrieved May 10, 2008, from https://www.cu.edu/ColemanInstitute/stateofth estates/UnitedStates_page3.pdf

Braddock, D., Hemp, R., & Rizzolo, M. (2008). *The state of the states in developmental disabilities 2008.* Denver, CO: University of Colorado, Department of Psychiatry and Coleman Institute for Cognitive Disabilities.

Bruininks, R. H., Woodcock, R. W., Weatherman, R. F., & Hill, B. K. (1996). *Scales of Independent Behavior-Revised.* Itasca, IL: Riverside.

Buunk, B. P., Doosje, B. J., Jans, L. G. J. M., & Hopstaken, L. E. M. (1993). Perceived reciprocity, social support, and stress at work: The role of exchange and communal orientation. *Journal of Personality and Social Psychology, 65,* 801–811.

Buunk, B. P., & Hoorens, V. (1992). Social support and stress. The role of social comparison and exchange processes. *British Journal of Clinical Psychology, 31,* 445–457.

Buunk, B. P., & Mutsaers, W. (1999). Equity perceptions and marital satisfaction in former and current marriages: A study among the remarried. *Journal of Social and Personal Relationships, 16,* 123–132.

Buunk, B. P., & Prins, K. S. (1998). Loneliness, exchange orientation, and reciprocity in friendships. *Personal Relationships, 5,* 1–14.

Carruth, A. K. (1996). Development and testing of the caregiver reciprocity scale. *Nursing Research, 45,* 93-100.

Carruth, A. K., Holland, C., & Larsen, L. (2000). Development and psychometric evaluation of the Caregiver Reciprocity Scale II. *Journal of Nursing Measurement, 8,* 179-91.

Carruth, A. K., Tate, U. S., Moffett, B. S., & Hill, K. (1997). Reciprocity, emotional well-being, and family functioning as determinants of family satisfaction in caregivers of elderly parents. *Nursing Research, 46,* 93-100.

Center, J., Beange, H., & McElduff, A. (1998). People with mental retardation have an increased risk for osteoporosis: A population study. *American Journal on Mental Retardation, 103,* 19–28.

Chen, S. C., Ryan-Henry, S., Heller, T., & Chen, E. (2001). Health status of mothers of adults with intellectual disability. *Journal of Intellectual Disability Research, 45,* 439-449.

Cheung, J., & Hicking, P. (2004). The experience of spousal carers of people with multiple sclerosis. *Qualitative Health Research, 14,* 153-66.

Cohen, J. (1988). *Statistical power analysis for the behavioral sciences (2nd ed.).* Hillsdale, NJ: Erlbaum.

Cook, K. (Ed) (1987). *Social exchange theory.* Newbury Park, CA: Sage.

Crocker, A.C. (2006). The developmental disabilities. In I. L. Rubin & A. C. Crocker (Eds.), *Medical care for children & adults with developmental disabilities.* (2nd ed., pp. 15-22). Baltimore, Maryland: Paul H Brookes Publishing.

Diener, E., & Emmons, R. (1984). The independence of positive and negative affect. *Journal of Personality and Social Psychology, 47,* 1105-17.

Dwyer, J. W., Lee, G. R., & Jankowski, T. B. (1994). Reciprocity, elder satisfaction and caregiver stress and burden: The exchange of aid in the family caregiving relationship. *Journal of Marriage and the Family, 56,* 35-43.

Dwyer, J. W., & Miller, M. K. (1990). Differences in caregiving network by area of residence: Implications for primary caregiver stress and burden. *Family Relations, 39,* 27-37.

Dumas, J., Wolf, L., Fisman, S., & Culligan, A. (1991). Parenting stress, child behavior problems, and dysphoria in parents of children with autism, Down syndrome, behavior disorders, and normal development. *Exceptionality, 2,* 97-110.

Dura, J. R., & Kiecolt-Glaser, J. K. (1990). Sample bias in caregiving research. *Journal of Gerontology: Psychological Sciences, 45,* P200-P204.

Emerson, E. (2001). *Challenging behaviour. Analysis and intervention in people with severe intellectual disabilities* (2nd ed.). Cambridge University Press, Cambridge.

Essex, E. L., Seltzer, M. M., & Krauss, M. W. (1999). Differences in coping effectiveness and well-being among aging mothers and fathers of adults with mental retardation. *American Journal on Mental Retardation, 104*, 545–563.

Evenhuis, H. M. (1995a). Medical aspects of ageing in a population with intellectual disability: I. Visual impairment. *Journal of Intellectual Disability Research, 39,* 19–26.

Evenhuis, H. M. (1995b). Medical aspects of ageing in a population with intellectual disability: II. Hearing impairment. *Journal of Intellectual Disability Research, 39,* 27–33.

Evenhuis, H. M. (1999). Associated medical aspects. In M. P. Janicki & A. J. Dalton (Eds.), *Dementia, aging and intellectual disabilities: A handbook* (pp. 103–118). Philadelphia: Brunner/Mazel.

Evenhuis, H., Henderson, C. M., Beange, H., Lennox, N., & Chicoine, B. (2001). Healthy ageing – adults with intellectual disabilities: Physical health issues. *Journal of Applied Research in Intellectual Disabilities, 14,* 175-194.

Finch, J., & Mason, J. (1993). *Negotiating family responsibilities.* London: Routledge.

Freedman, R. I., Krauss, M. W., & Seltzer, M. M. (1997). Aging parents' residential plans for adult children with mental retardation. *Mental Retardation, 35,* 114–123.

Folkman, S. (1997). Positive psychological states and coping with severe stress. *Social Science & Medicine, 45,* 1207-1221.

Folkman, S., Chesney, M. A., Cooke, M., Boccellari, A, & Collette, L. (1994). Caregiver burden in HIV+ and HIV- partners of men with AIDS. *Journal of Consulting and Clinical Psychology, 62,* 746-756.

Folkman, S., & Lazarus, R. S. (1980). An analysis of coping in a middle-aged community sample. *Journal of Health and Social Behavior, 21,* 219–239.

Fujiura, G. T. (1998). Demography of family households. *American Journal on Mental Retardation, 103,* 225-235.

Fujiura, G. T., Roccoforte, J. A., & Braddock, D. (1994). Costs of family care for adults with mental retardation and related developmental disabilities. *American Journal on Mental Retardation, 99,* 250-261.

General Accounting Office. (1999, June). *SSI children: Multiple factors affect families' costs for disability-related services, GAO/HEHS-99-99.* Washington DC: Author.

Gouldner, A. W. (1973). The norm of reciprocity: A preliminary statement. In A. W. Gouldner (Ed.), *For sociology* (pp. 226-259). London: Allan Lane.

Grant, G. (1990). Elderly parents with handicapped children: Anticipating the future. *Journal of Aging Studies, 4*, 359–374.

Grant. G., & McGrath, M. (1990). Need for respite-care services for caregivers of persons with mental retardation. *American Journal on Mental Retardation, 94*, 638-648.

Grant, G., Ramcharan, P., McGrath, M., Nolan, M., & Keady, J. (1998). Rewards and gratifications among family caregivers: towards a refined model of caring and coping. *Journal of Intellectual Disability Research, 42*, 58–71.

Haag, H., Ruther E., & Hippius, H. (1992). *Tardive Dyskinesia. WHO Expert Series on Biological Psychiatry.* Seattle, WA: Hogrefe & Huber.

Haley, W. E., & Bailey, S. (1999). Research on family caregiving in Alzheimer's disease: Implications for practice and policy. In B. Vellas & J. L. Fitten (Eds.), *Research and practice in Alzheimer's disease: Vol. 2* (pp. 321-332). Paris: Serdi.

Haley, W. E., Levine, E. G., Brown, S. L., & Bartolucci, A. A. (1987). Stress, appraisal and social support as predictors of adaptational outcomes among dementia caregivers. *Psychology and Aging, 2*, 323-330.

Haley, W. E., & Perkins, E. A. (2004). Current status and future directions in family caregiving and aging people with intellectual disability. *Journal of Policy and Practice in Intellectual Disabilities, 1*, 24-30.

Harwood, D. G., Ownby, R. L., Burnett, K., Barker, W. W., & Duara, R. (2000). Predictors of appraisal and psychological well-being in Alzheimer's disease family caregivers. *Journal of Clinical Geropsychology, 6*(4), 279-297.

Hayden, M. F., & Heller, T. (1997). Support, problem-solving/coping ability, and personal burden of younger and older caregivers of adults with mental retardation. *Mental Retardation, 35*, 364-72

Heaton-Ward, A. (1975). *Mental subnormality* (4th ed.). Bristol: John Wright.

Heller, T., & Factor, A. (1993). Aging family caregivers: Support resources and changes in burden and placement desire. *American Journal on Mental Retardation, 98*, 417-426.

Heller, T., Miller, A. B., & Factor, A. (1997). Adults with mental retardation as supports to their parents: Effects on parental caregiving appraisal. *Mental Retardation, 35*, 338–346.

Heller, T., Rowitz, L., & Farber, B. (1992). *The domestic cycle of families of persons with mental retardation* (Rep.). Chicago, IL: University of Illinois at Chicago, Affiliated Program in Developmental Disabilities and School of Public Health.

Hirschfeld, M. (1983). Home care versus institutionalisation: Family caregiving and senile brain disease. *International Journal of Nursing Studies, 20,* 23 – 32.

Horwitz, A. V., Reinhard, S. C., & Howell-White, S. (1996). Caregiving as reciprocal exchange in families with seriously mentally ill members. *Journal of Health and Social Behaviour, 37,* 149-162.

Jacobson, J. & Mulick, J. A. (Eds.). (1996). *Manual of Diagnosis and Professional Practice in Mental Retardation.* Washington, DC: American Psychological Association.

Janicki, M. P. (1996, Fall). Longevity increasing among older adults with an intellectual disability. *Aging, Health, & Society, 2, 2.*

Janicki, M. P., & Dalton, A. J. (2000). Prevalence of dementia and impact on intellectual disability services. *Mental Retardation, 38,* 276–288.

Janicki, M. P., Dalton, A. J., Henderson, C. M., & Davidson, P. W. (1999). Mortality and morbidity among older adults with intellectual disability: Health services considerations. *Disability and Rehabilitation, 21,* 284-294.

Jung, J. (1990). The role of reciprocity in social support. *Basic and Applied Social Psychology, 11,* 243-253

Katz, S., Ford, A.B., & Moskowitz, R.W. (1963). The index of ADL: A standardized measure of biological and psychosocial function. *Journal of the American Medical Association, 185,* 914–919.

Kawachi, I., Kennedy, B. P., & Glass, R. (1999). Social capital and self-rated health: A contextual analysis. *American Journal of Public Health, 89,* 1187–1193.

Kelly, T.,& Kropf, N. P. (1995). Stigmatized and perpetual parents: older parents caring for adult children with life-long disabilities. *Journal of Gerontological Social Work, 24,* 3–16.

Kesey, K. (1962). *One flew over the cuckoo's nest.* New York: Penguin.

Kiecolt-Glaser, J. K., Dura, J. R., Speicher, C.E., Trask, O. J., & Glaser, R. (1991). Spousal caregivers of dementia victims: Longitudinal changes in immunity and health. *Psychosomatic Medicine, 53,* 345-362.

Kiecolt-Glaser, J. K., Marucha, P. T., Malarkey, W. B., Mercado, A. M., & Glaser, R. (1995). Slowing of wound healing by psychological stress. *The Lancet, 346,* 1194-1196.

King, A. C., Oka, R. K., & Young, D. R. (1994). Ambulatory blood pressure and heart rate responses to the stress of work and caregiving in older women. *Journal of Gerontology: Medical Sciences, 49,* M239-M245.

Krause, N. (1995). Negative interaction and satisfaction with social support among older adults. *Journal of Gerontology: Psychological Sciences, 50B,* P59-P73.

Krause, N., & Borawski-Clark, E. (1995). Social class differences in social support among older adults. *The Gerontologist, 35,* 498-508.

Lakin, K. C., Prouty, R., & Coucouvanis, K. (2006). Changing patterns in size of residential settings for persons with intellectual and developmental disability, 1997-2005. *Mental Retardation, 44,* 306-309.

Langa, K. M., Chernew, M. E., Kabeto, M. U., Herzog, A. R., Ofstedal, M. B., Willis R. J., et al. (2001). National estimates of the quantity and cost of informal caregiving for the elderly with dementia. *Journal of General Internal Medicine, 16,* 770-778.

Lawton, M. P. (1983). Environmental and other determinants of well-being in older people. *The Gerontologist, 23,* 349–57.

Lawton, M.P., & Brody, E. M. (1969). Assessment of older people: Self-maintaining and instrumental activities of daily living. *The Gerontologist, 9,* 179-186.

Lazarus, R. S., & Folkman, S. (1984). *Stress, appraisal, and coping.* New York: Springer.

Lee, S., Colditz, G. A., Berkman, L., & Kawachi, I. (2003). Caregiving and risk of coronary heart disease in U.S. women: a prospective study. *American Journal of Preventive Medicine, 24,* 113-119.

Levi-Straus, C. (1964). Reciprocity, the essence of social life. In R. L. Coser (Ed.), *The family: Its structure and functions* (pp. 36-48). New York: St. Martin's Press.

Loomis, L. S., & Booth, A. (1995). Multigenerational caregiving and well-being: The myth of the beleaguered sandwich generation. *Journal of Family Issues, 16,* 131-148.

Martínez-Martín, P., Benito-León, J., Alonso, F., Catalán, M. J., Pondal' M., Zamarbide, I., et al. (2005). Quality of life of caregivers in Parkinson's disease. *Quality of Life Research, 14,* 463-472.

McCullagh, E., Brigstocke, G., Donaldson, N., & Kalra, L. (2005). Determinants of caregiving burden and quality of life in caregivers of stroke patients. *Stroke, 36,* 2181-2186.

McDonell, M.G., Short, R. A., Berry, C. B., & Dyck, D. G. (2003). Burden in schizophrenia caregivers: Impact of family psychoeducation and awareness of patient suicidality. *Family Process, 42,* 91-103.

Molm, L. D., & Cook, K. S. (1995). Social exchange and exchange networks. In K. S. Cook, G. A. Fine, & J. S. House (Eds.), *Sociological perspectives on social psychology* (pp.209-235). Needham Heights, MA: Allyn & Bacon.

Morycz, R. K. (1985). Caregiving strain and the desire to institutionalize family members with Alzheimer's disease: Possible predictors and model development. *Research on Aging, 7,* 329-361.

National Alliance for Caregiving & AARP. (2004). *Caregiving in the US.* Retrieved May 9, 2006 from http://www.caregiving.org/data/04finalreport.pdf

National Family Caregivers Association & Family Caregiver Alliance. (2006). *Prevalence, hours and economic value of family caregiving, updated state-by-state analysis of 2004 national estimates* (by Peter S. Arno, PhD). Kensington, MD: NFCA & San Francisco, CA: FCA

Neufeld, A., & Harrison, M. J. (1995). Reciprocity and social support in caregiver's relationships: Variations and consequences. *Qualitative Health Research, 5,* 348-365.

Nijboer, C., Triemstra, M., Tempelaar, R., Sanderman, R., & van den Bos, G. A. (1999). Determinants of caregiving experiences and mental health of partners of cancer patients. *Cancer, 86,* 577-588.

Parish, S. L., Rose, R. A., Grinstein-Weiss, M., Richman, E. L., & Andrews, M. E. (2008). Material hardship in U. S. families raising children with disabilities. *Exceptional Children, 75,* 71-92.

Parish, S. L., Seltzer, M. M., Greenberg, J. S., & Floyd, F. (2004). Economic implications of caregiving at Midlife: Comparing parents with and without children who have developmental disabilities. *Mental Retardation, 42,* 413-426.

Pearlin, L. I., Mullan, J. T., Semple, S. J., & Skaff, M. M. (1990). Caregiving and the stress process: An overview of concepts and their measures. *The Gerontologist, 30,* 583-594.

Pearlin, L., & Schooler, C. (1978). The structure of coping. *Journal of Health and Social Behavior, 24,* 2–15.

Perkins, E. A. (in press). The compound caregiver: A case study of multiple caregiving roles. *Clinical Gerontologist.*

Perkins, E. A., Lynn, N., & Haley, W. E. (2007). Caregiver issues associated with wandering. In A. Nelson, & D. Algase (Eds.), *Evidence-based protocols for managing wandering behaviors* (pp. 123-141). New York: Springer.

Phillips, J. (1998). Complications of anticonvulsant drugs and ketogenic diet. In J. Biller (Ed.), *Iatrogenic Neurology* (pp. 397–414). London: Butterworth-Heinemann.

Pinquart, M., & Sörensen, S. (2003). Differences between caregivers and noncaregivers in psychological health and physical health: a meta-analysis. *Psychology and Aging, 2,* 250- 267.

Prouty, R., Smith, G., & Lakin, K. C. (2003). *Residential services for persons with developmental disabilities: Status and trends through 2002.* Minneapolis: University of Minneapolis, Research and Training Center on Community Living/Institute on Community Intergration.

Pruchno, R., Burant, C., & Peters, N. (1997). Understanding the well-being of care receivers. *The Gerontologist, 37,* 102–109.

Radloff, L. (1977). The CES-D Scale: a self-report depression scale for research in the general population. *Applied Psychological Measurement, 1,* 385-401.

Reid, C. E., Moss, S., & Hyman, G. (2005). Caregiver reciprocity: The effect of reciprocity, carer self-esteem and motivation on the experience of caregiver burden. *Australian Journal of Psychology, 57,* 186-196.

Robinson-Whelen, S. R., Tada, Y., MacCallum, R. C., McGuire, L., & Kiecolt-Glaser, J. K. (2001). Long-term caregiving: What happens when it ends? *Journal of Abnormal Psychology, 110,* 573-584.

Rogerson, P. A., & Kim, D. (2005). Population distribution and redistribution of the baby-boom cohort in the United States: Recent trends and implications. *Proceedings of the National Academy of Sciences, 102,* 15319-15324.

Rook, K. (1987). Reciprocity of social exchange and social satisfaction among older women. *Journal of Personality and Social Psychology, 52,* 145 – 154.

Sahlins, M. D. (1965). On the sociology of primitive exchange. In M. Banton (Ed.), *The Relevance of models for social anthropology* (pp. 159-237). London: Tavistock.

Schrojenstein Lantman-de Valk, H. M. J., van den Akker, M., Maaskant, M. A., Haveman, M. J., Urlings, H. F. J., Kessels, A. G. H., et al. (1997). Prevalence and incidence of health problems in people with intellectual disability. *Journal of Intellectual Disability Research, 41,* 42-51.

Schrojenstein Lantman-de Valk, H. M. J., Haveman, M. J., Maaskant, M. A., & Kessells, A. G. (1994). The need for assessment of sensory functioning in ageing people with mental handicap. *Journal of Intellectual Disability Research, 38,* 289-298.

Schulz, R., & Beach, S. R. (1999). Caregiving as a risk for mortality. The caregiver health effects study. *Journal of the American Medical Association, 282,* 2215-2219.

Schulz, R., Mendelsohn, A. B., Haley, W. E., Mahoney, D., Allen, R. S., Zhang, S., et al. 2003). End-of-life care and the effects of bereavement on family caregivers of persons with dementia. *The New England Journal of Medicine, 349,* 1936-1942.

Schulz, R., O'Brien, A. T., Bookwala, J., & Fleissner, K. (1995). Psychiatric and physical morbidity effects of dementia caregiving: Prevalence, correlates, and causes. *The Gerontologist, 35,* 771-791.

Selikowitz, M. (1990). *Down syndrome: The facts.* New York: Oxford University Press.
Seltzer, M. M., Greenberg, J. S., Floyd, F. J., Pettee, Y., & Hong, J. (2001). Life course impacts of parenting a child with a disability. *American Journal on Mental Retardation, 106,* 265-286.

Sörensen, S., Pinquart, M., & Duberstein, P. (2002). How effective are interventions with caregivers? An updated meta-analysis. *The Gerontologist, 42,* 356-372.

Suitor, J. J., Pillemer, K., & Sechrist, J. (2006). Within-family differences in mothers' support to adult children. *Journal of Gerontology: Social Sciences, 61B,* S10–S17.

Tabachnick, B. G., & Fidell, L. S. (1996). *Using multivariate statistics* (3rd ed.). New York: Harper Collins

Tarlow, B. J., Wisniewski, S. R., Belle, S. H., Rubert, M., Ory, M. G., & Gallagher-Thompson, D. (2004). Positive aspects of caregiving: contributions of the REACH project to the development of new measures for Alzheimer's caregiving. *Research on Aging*, 26, 429- 453.

Thompson, L. (2004, March). Long Term Care Financing Project, Georgetown University. Issue Brief. *Long-term care: support for family caregivers*. Retrieved May 8, 2006 from http://ltc.georgetown.edu/pdfs/caregivers.pdf

Todd, S., & Shearn, J. (1996). Time and the person: Impact of support services on the lives of parents of adults with intellectual disabilities. *Journal of Applied Research in Intellectual Disabilities, 9,* 40-60.

Townsend, A., Noelker, L., Deimling, G., & Bass, D. (1989). Longitudinal impact of inter-household caregiving on adult children's mental health. *Psychology and Aging, 4,* 393-401.

United Nations. (1971). *UN declaration of the rights of the mentally retarded.* Retrieved August 10, 2007, from http://www.unhchr.ch/html/menu3/b/m_mental.htm

United.States Census Bureau. (2007). *Table 1: Estimates of the population by selected age groups for the United States and states for Puerto Rico, July 1, 2006 (SC-EST2006-01).* Washington, DC: Author.

Väänänen, A., Buunk, B. P., Kivimäki, M., Pentti, J., & Vahtera, J. (2005). When is it better to give than to receive: Long-term health effects of perceived reciprocity in support exchange. *Journal of Personality and Social Psychology, 89,* 76-193.

Vanderwerker, L. C., Laff, R. E., Kadan-Lottick, N. S., McColl, S., & Prigerson, H. G. (2005). Psychiatric disorders and mental health service use among caregivers of advanced cancer patients. *Journal of Clinical Oncology, 23,* 6899-6907.

Vitaliano, P. P., Scanlan, J. M., Krenz, C., Schwartz, R. S., & Marcovina, S. M., (1996). Psychological distress, caregiving, and metabolic variables. *Journal of Gerontology:Psychological Sciences, 51,* P290-P299.

Vitaliano, P. P., Zhang, J., & Scanlan, J. M. (2003). Is caregiving hazardous to one's physical health? A meta-analysis. *Psychological Bulletin, 129,* 946-972.

Walden, S., Pistrang, N., & Joyce, T. (2000). Parents of adults with intellectual disabilities: Quality of life and experiences of caring. *Journal of Applied Research in Intellectual Disabilities, 13,* 62-76.

Walker, A. J., & Pratt, C. C. (1991). Daughters' help to mothers: Intergenerational aid versus caregiving. *Journal of Marriage and the Family, 53,* 3-12.

Walster, E., Walster, G., & Bersheid, E. (1978). *Equity theory and research.* Boston: Allyn & Bacon.

Warburg, M. (1994). Visual impairment among people with developmental delay. *Journal of Intellectual Disability Research, 38,* 423-432.

Ware, J. E., Jr., Kosinski, M., Bjorner, J. B., Turner-Bowker, D. M., Gandek, B., & Maruish, M. E. (2007). *User's manual for the SF-36v2™ Health Survey* (2nd. Ed.) Lincoln, RI: Quality Metric Incorporated.

Ware, J. E., & Sherbourne, C. D. (1992). The MOS 36-item short-form health survey (SF-36). I. Conceptual framework and item selection. *Medical Care, 30,* 473-83.

Watson, D., & Tellegen, A. (1985). Toward a consensual structure of mood. *Psychological Bulletin, 98,* 219-35.

Wilson, D. N., & Haire, A. (1990). Health care screening for people with mental handicap living in the community. *British Medical Journal, 301,* 1379-1381.

Wojcieszek, J. (1998). Drug-induced movement disorders. In J. Biller (Ed.), *Iatrogenic Neurology* (pp. 215-230). London: Butterworth-Heinemann.

Wolfensberger, W. (1972). *The principle of normalization in human services.* Toronto: National Institute on Mental Retardation.

Wolfensberger, W. (1976). The origin and nature of our institutional models. In R. Kugel & A. Shearer (Eds.), *Changing patterns in residential services for the mentally retarded* (Rev.ed., pp. 35-82). Washington, DC: President's Committee on Mental Retardation.

Wolfensberger, W. (1983). Social role valorization: A proposed new term for the principle of normalization. *Mental Retardation, 12,* 231-239.

Wolff, J. L., & and Agree E. M. (2004). Depression among recipients of informal care: The effects of reciprocity, respect, and adequacy of support. *Journal of Gerontology: Social Sciences 59B,* S173–S180.

Wood, V., Wylie, M. L., & Schaefor, B. (1969). An analysis of a short self-report measure of life satisfaction: correlation with rater judgments. *Journal of Gerontology, 24,* 465-469.

World Health Organization. (2001). *World Health Report 2001 – Mental Health: New Understanding, New Hope.* Retrieved March 26, 2008, from http://www.who.int/whr/ 2001/en/whr01_ch2_en.pdf

Zarit, S. H., Reever, K. E., & Bach-Peterson, J. (1980). Relatives of the impaired elderly: Correlates of feelings of burden. *The Gerontologist, 20,* 649-655.

About the Author

Elizabeth A. Perkins is a Registered Nurse Mental Handicap (RNMH) from the Hereford and Worcester College of Nursing and Midwifery, England from 1992. She also has a Bachelor's Degree (Summa Cum Laude) in Psychology from the University of South Florida from 2003. She previously worked in clinical and managerial roles in residential care of both older adults and persons with intellectual/developmental disabilities in England. She entered the Ph.D. in Aging Studies program at the University of South Florida in 2003.

While in the Ph.D. program at the University of South Florida, Ms. Perkins was employed as a Graduate Teaching Associate, teaching undergraduate Physical Changes and Aging, and Psychology of Aging courses. She co-authored five peer-reviewed journal articles, one book chapter, and one textbook. She has presented her work at several national conferences including the Gerontological Society of America, and the American Association on Intellectual and Developmental Disabilities.

CPSIA information can be obtained
at www.ICGtesting.com
Printed in the USA
LVIC070509280313
326403LV00005B